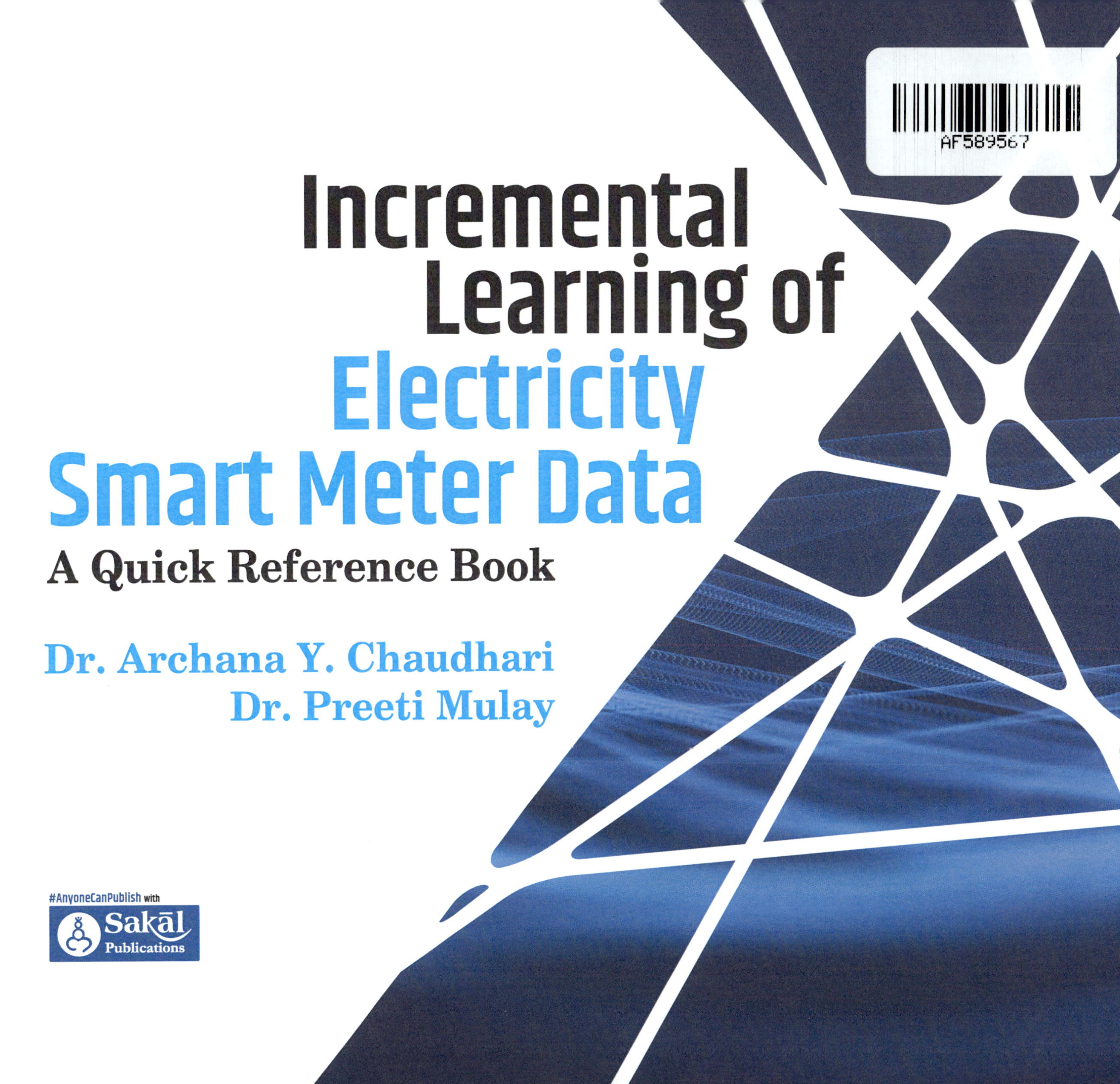

AF589567

Incremental Learning of Electricity Smart Meter Data

A Quick Reference Book

Dr. Archana Y. Chaudhari
Dr. Preeti Mulay

#AnyoneCanPublish with
Sakāl Publications

#AnyoneCanPublish with

Incremental Learning of Electricity Smart Meter Data

© Dr. Archana Y. Chaudhari, Dr. Preeti Mulay, 2023

Authors : Dr. Archana Y. Chaudhari, Dr. Preeti Mulay
First Edition : February 2023
Publisher : Sakal Media Pvt Ltd.
595, Budhwar Peth,
Pune 411 002
Cover design and Layout : Yashodhan Lovalekar, Pune

ISBN : 978-93-95139-52-6

For Details : 020-24405678 / 88888 49050
sakalprakashan@esakal.com

© All rights reserved.

No part of this publication may be reproduced or transmitted in any form or by any means, electronically or mechanically, including photocopying, recording, broadcasting, pod casting of any information storage or retrieval system without prior permission in writing form the writer or in accordance with the provisions of the Copy Right Act (1956) (as amended). Any person who does any unauthorised act in relation to this publication may be liable to criminal prosecution and civil claims for damages.

Disclaimer :

Although the author has taken every effort to ensure that the information in this book was correct at the time of printing, the author and publisher do not assume and hereby disclaim any liability to any party, society for any loss, damage, or disruption caused by errors or omissions, whether such errors and omissions are caused due to negligence, accident, amendment in Act, Rules, Bye laws or any other cause. The views expressed in this book are those of the Authors and do not necessarily reflect the views of the Publishers.

A goal is not always meant to be reached;
it often serves simply as something to aim at.

- Bruce Lee

Preface

According to the World Resources Institute, energy-related emissions make up more than two-thirds of India's overall emissions and represent more than three times the next largest source (the industry sector). Of the country's energy sector emissions, 77% are from electricity generation, making this a key target for reductions to meet India's climate commitments. Although India's emissions are still comparatively low on a per-capita basis, the very size of the nation's population and the scope for them to increase make the country's emissions a global concern. India's emissions from the energy sector are the fourth largest in the world (behind China, the United States, and the European Union), and rising.

One of the challenges faced by India's electricity sector is the capacity of the grid to accommodate renewable energy as the country invests in it to meet national targets. It will have a potential surplus, rather than a shortfall, in electricity; yet it faces technical constraints in using that electricity. According to analyst firm Brookings, to avoid curtailing renewable energy generation, the country needs a set of stronger grids, cheap storage, and the ability to shift load to match supply conditions.

In parallel with these challenges, utility companies in India still rely on manual electricity meter reading to track consumption at residential and industrial locations. Local utility companies send employees to read these meters each month and then generate monthly electricity bills. Generation of data regarding consumption patterns compiled from manual readings is prone to errors, time consuming and not as widely comprehensive as may be needed to arrive at planning for providing sufficient energy at maximum efficiency and economical cost. This system does not provide utilities with insight into electricity consumption patterns, for example, based on time-of-day usage, nor does it enable consumers to understand their electricity consumption as part of the larger picture of electricity availability.

Electricity Smart Meters (ESMs) with a machine-learning layer on top will not only allow utility companies to monitor electricity usage in real time but also streamline the process of electricity distribution and cut carbon emissions. ESMs are an integral part of a flexible, decentralized and decarbonized energy system. ESM has inbuilt electronics such as embedded software to record and transmit customer electricity consumption to utilities in specific time intervals of an hour or less. In addition to saving the laborious process of manual meter readings, ESM enables utilities to monitor energy usage in real time.

Furthermore, ESM generates a huge amount of data incrementally, which become the source for analytics geared towards improved overall energy management and emission reduction. However, on an influx of new data, traditional clustering tasks re-cluster all of the data from scratch. The incremental clustering method is essential to solve the problem of grouping growing data for pattern recognition. In order to reduce carbon emissions from the production of electricity, the exact consumption pattern of the consumer is required. It is most essential to keep learning incrementally for ever-increasing consumption data for effectual decisions, predictions and problem solving. This can be achieved effectually by applying incremental clustering methods on real ESM data sources. Incremental learning can be achieved via incremental clustering easily as well as effectively.

The present research considers not only residential electricity consumption data but also related socioeconomic data such as the variability of weather conditions at meter installation sites, related geographic information and demographic data. However, given the volume of data and the number of data types involved, ESM data analytics are highly complex. To achieve worldwide data analysis of ESM data and broader perspectives, it is essential to deploy incremental clustering algorithms on distributed platforms, which will enable them to accept ESM data from varied sources, analyse it and produce distributed worldwide solutions. Microsoft Azure provides the processing power necessary to handle such dynamic systems and perform complex analytics jobs.

Hence, this research focuses on ESM data analysis by proposed Cloud for Closeness-based Gaussian Mixture Incremental Clustering Algorithm (Cloud4CGMIC). The proposed system has been validated on real-world ESMs datasets like Irish Social Science Data Archive (Ireland) dataset, Low Carbon London dataset and Prayas Energy Group (India) dataset, to name a few. The proposed system can capture and generate the hidden pattern of consumption based on day-time and night-time, season-wise, and area-specific learning.

The distributed incremental clustering system identifies the change in residential energy consumption and provides valuable data to utility companies to optimize electricity load management. Also, utility providers can improve security, reduce peak loads, increase integration of renewable energy, and reduce capital expenses of building new plants. Its greater transparency and responsive information will allow household customers to better manage their energy consumption and reduce electricity bills. These analysed details are useful for a generation unit to reduce electricity generation in peak times and, possibly, emissions, facilitate balance supply and demand, and increase short-term reliability of supply. This research will be extremely helpful for the environment by reducing pollution via carbon production by power plants (as pollution is hazardous to health). This invention thus assists the government in efficiently planning for the economic and energy needs of the growing population.

Acknowledgement

- It gives us great pleasure and immense satisfaction to express our deepest sense of gratitude to everyone who has directly or indirectly helped us in completing this "Quick Reference Book" successfully.
- We express our gratitude towards Symbiosis Institute of Technology, Symbiosis International (Deemed University), Pune, India for providing the wings to broaden the research orientation and conduct innovative work in a supportive environment always.
- We would like to thank Director, Symbiosis Institute of Technology and Dean, Faculty of Engineering, Symbiosis International (Deemed University), Pune, India for allowing us to live with our dream in this institute and also encouraging, mentoring and motivating us consistently in all aspects.
- We would like to thank research in-charge, Deputy Director (Academics), Deputy Director (Administration), all the HoDs, faculty colleagues, and all our staff friends.
- We also express our deep gratitude towards Symbiosis Center for Research and Innovation (SCRI) for the guidance in our overall research journey, papers, patents and now this book publication.
- We are also thankful to all the anonymous reviewers of our papers that helped us improve our research and recognized our contribution to the existing knowledge base.
- We gratefully acknowledge the funding agency, Microsoft Azure: AI for Earth and Sakal India Foundation for providing cloud resources and financial support, respectively, to complete our thesis work.
- No words are sufficient to express our gratitude towards our family and friends for their kind support and encouragement. Further, without the support and help from our publisher, it would not have been possible for us to pursue and achieve the ultimate aim of publishing this book.
- Last but not least, we would also like to extend our appreciation to those who are not mentioned here but have played a crucial role and inspired this work.

Table of Contents

List of Figures

List of Tables

List of Abbreviations

AI	Artificial Intelligence
AML	Advanced Machine Learning
ANN	Artificial Neural Network
CBICA	Correlation Based Incremental Clustering Algorithm
CBTs	Customer Behavior Trials
CER	Commission for Energy Regulation
CFBA	Closeness Factor-Based Algorithm
CGMIC	Closeness based Gaussian Mixture Model Incremental Clustering
Cloud4CGMIC	Cloud for Closeness based Gaussian Mixture Model Incremental Clustering
DR	Demand Response
DBI	Davies-Bouldin Index
DCGMIC	Distributed CGMIC
DBSCAN	Density-Based Spatial Clustering of Applications with Noise
DI	Dunn Indicator
DisIClus	Distributed Incremental Clustering
ESM	Electricity Smart Meter
ESMDA	Electricity Smart Meter Data Analysis
EM	Expectation Maximization
FCM	Fuzzy C-Means
GMM	Gaussian Mixture Model
IC	Incremental Clustering
IL	Incremental Learning
IoT	Internet of Things
IT	Information Technology
LTLF	Long-Term Load Forecasting
LSTM	Long Short Term Memory
LCL	Low Carbon London
ML	Machine Learning
MTLF	Medium-Term Load Forecasting
MSE	Mean Squared Error
MAE	Mean Absolute Error
RNN	Recurrent Neural Network
SG	Smart Grid
VMs	Virtual Machines
WoS	Web of Science

Book Outline

- **Chapter 1: Overview of Smart Meter Data Analysis and Incremental Clustering:** Briefly discusses the incremental clustering methods and challenges in the learning methods, and identifies the need for incremental learning. Describes the process of a typical incremental learning method. Also highlights the applications of Electricity Smart Meter Data Analysis and Distributed Clustering.
- **Chapter 2: Problem Identification:** After extensive world-wide literature survey on existing or similar systems, the strengths and weaknesses of other research work were evaluated. Finally, based on the research gaps, we designed the novel concept termed as Electricity Smart Meter Data Analysis using incremental clustering algorithms. The objectives of the research are also highlighted in this chapter.
- **Chapter 3: Electricity Smart Meter Data Analysis using Incremental Clustering Algorithm:** Describes a method for improving energy management, enabling reduced electricity consumption and load profile based on analysis of ESM data. The disclosed invention redesigns and develops a Closeness-based Gaussian Mixture Incremental Clustering (CGMIC) Algorithm via a cloud platform (Microsoft Azure) for mining the hidden patterns of electrical energy consumption such as load composition, season-wise electrical usage patterns, time (day/night) specific patterns, etc.

  The developed scalable platform, called Cloud for CGMIC (Cloud4CGMIC), efficiently handled the incremental data and extracted the electricity consumption patterns of consumers much faster through distributed computations on many other machines.
- **Chapter 4: Load Forecasting using LSTM Network -** Describes LSTM-based load forecasting. The memory cell of the LSTM model learns useful information from historical data for a long period and forgets useless information. The performance of the LSTM model was validated by experimentally with real-world electricity load datasets from Toronto.
- **Chapter 5: Concluding Remarks and Future Research Directions -** Discusses the significant contribution of the work based on the various experiments carried out. This chapter also highlights the major contributions of the work and suggests the future scope.

Chapter 1

Overview of Smart Meter Data Analysis and Incremental Clustering

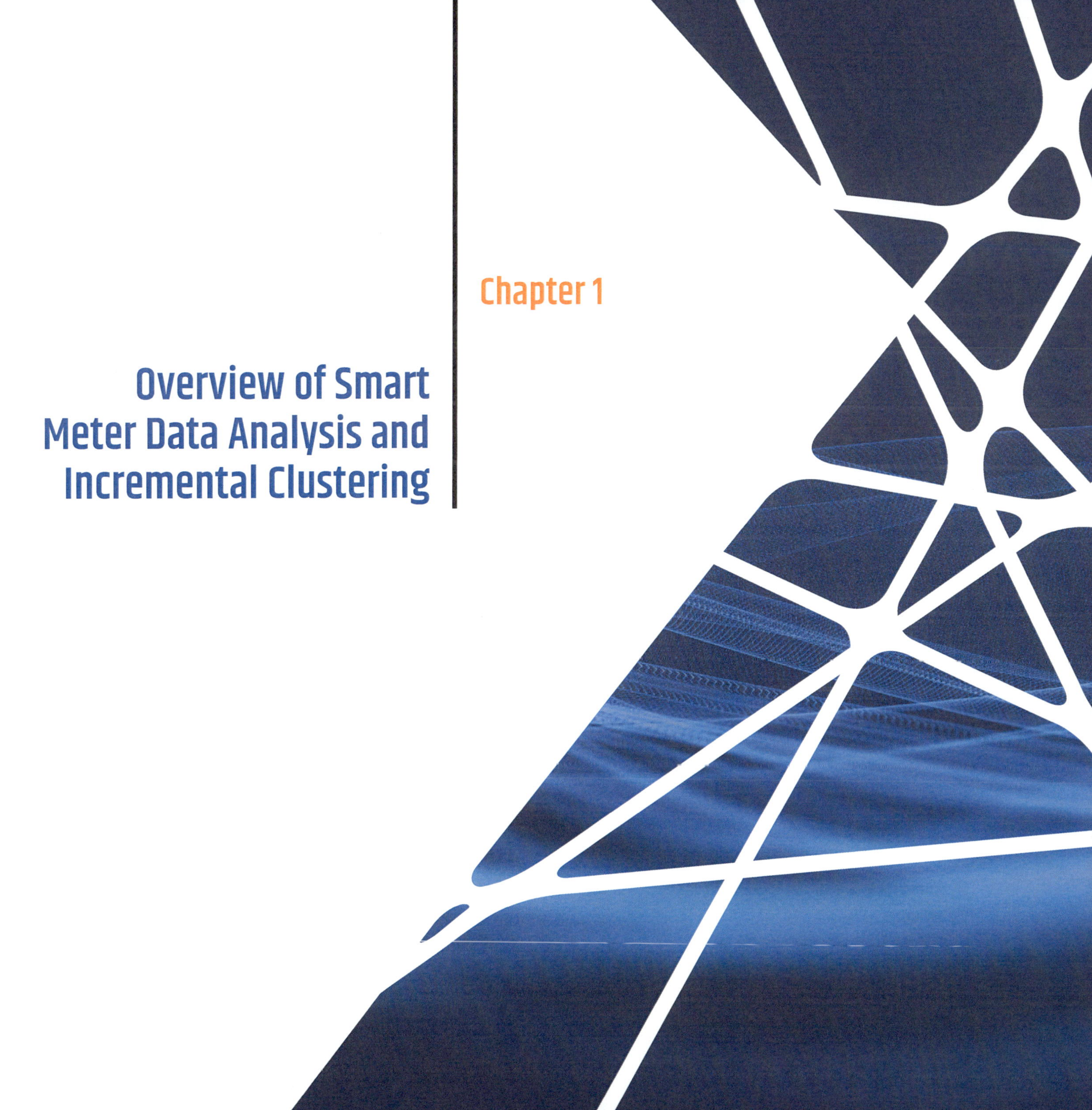

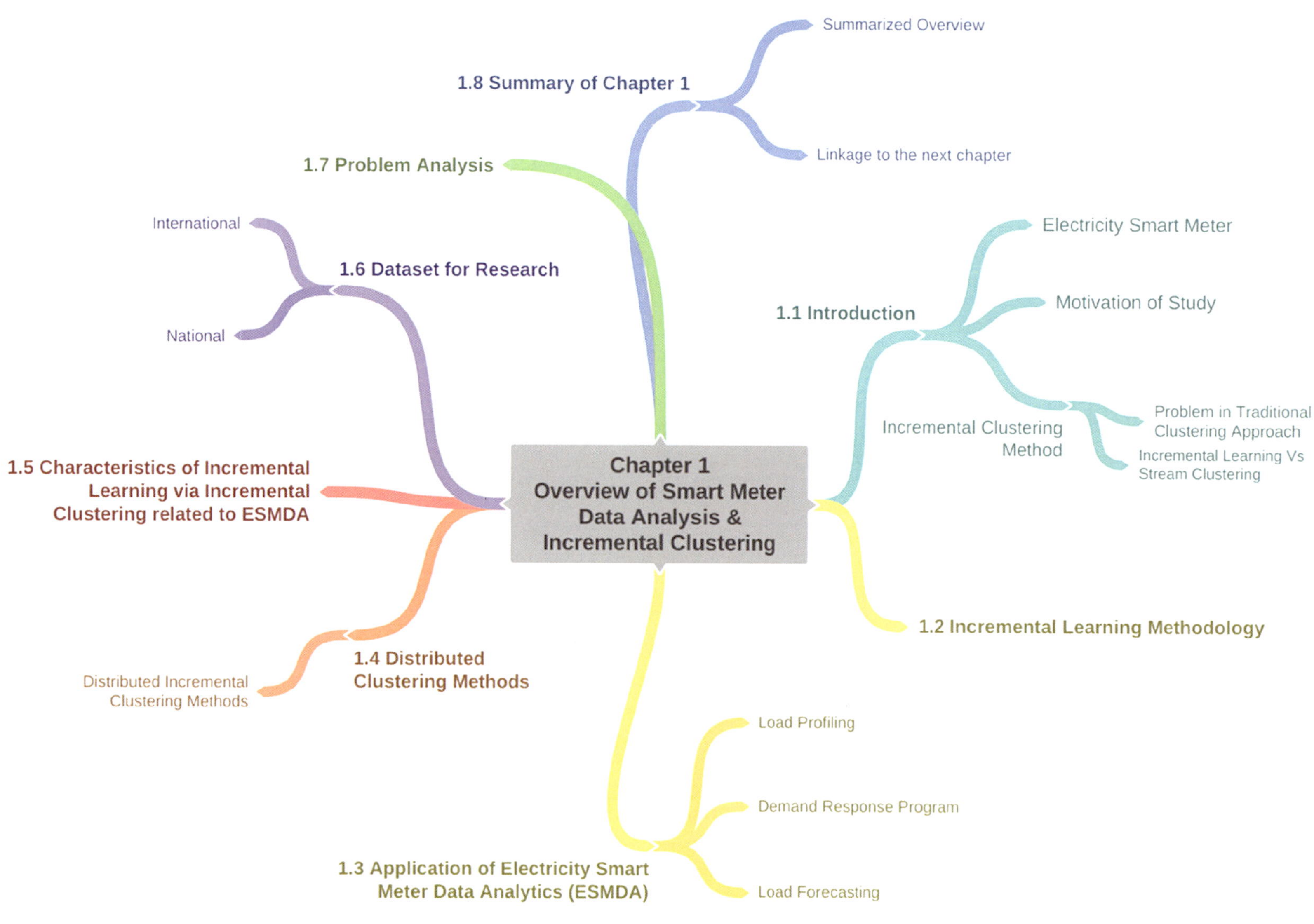

Mind map for the overall chapter 1

Source: https://coggle.it/

1.1 Introduction

- The Smart Grid (SG) is the next generation of the electrical grid that can deliver electricity in a smart, controlled way from the power plant to your home or business (Siano, 2014).
- SG allows two-way communication between the utility and its customers, and the sensing along the transmission lines makes the grid smart.
- SG permits consumers to modify their purchasing patterns and behavior according to the received information, incentives and disincentives (Li and Yao, 2010; EPRI, 2020).
- The advantages of SG are:
  - Detailed tracking of energy consumption,
  - Easily comprehensible description of anomalies,
  - Quicker restoration of electricity after power disruptions,
  - Saved costs of installing new utility base stations.
- Figure 1.1 has the size of the words scaled relatively to the maximum and minimum word frequency according to the available literature. Words in smaller fonts show very least research is carried out in those fields.

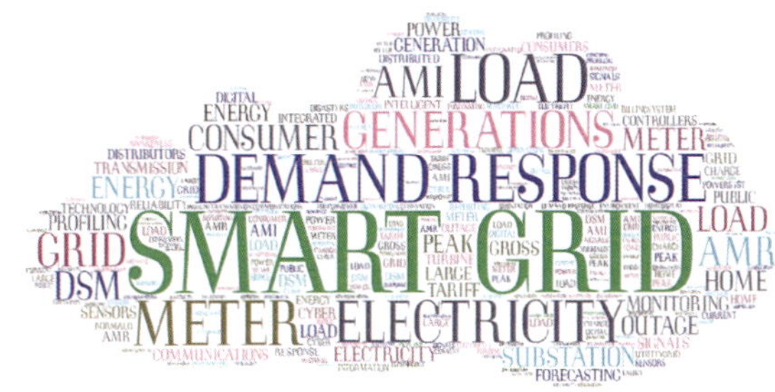

Fig. 1.1. Word Cloud for Smart Grid Concepts

- Effective functioning of SG requires sensors, communication systems, smart devices, specialized processors, and new technologies and equipment working together. Electricity Smart Meter (ESM) is one of the key technologies implemented in SG.

1.1.1 Electricity Smart Meter

- Electricity Smart Meters (ESMs) record the fine-grained energy consumption (electricity load) of customers and provide the recorded information to utility companies for advanced measurement and control applications as shown in Fig. 1.2.
- Compared to a conventional electricity meter (MSEB, 2018), an ESM has controls, automation and communication units to give consumers better information and automatically report outages (Yildiz et al., 2017).
- As the deployment of ESM is increasing, it generates a wealth of fine-grained incremental data to provide benefits to various stakeholders of power systems (customers, generation units and transmission and distribution units).
- However, such datasets are not useful without analytical powers.
- Analytics solutions will be able to obtain valuable insights into the huge data generated by ESM.

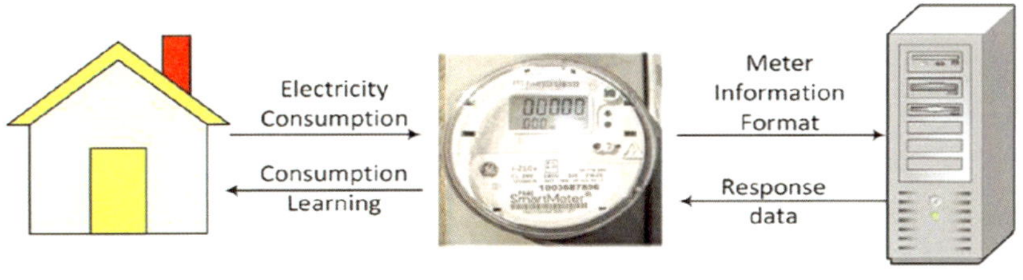

Fig. 1.2. Electricity Smart Meter Working in Real Time Environment (Source: Chaudhari and Mulay, 2020a)

1.1.2 Motivation of Study

- In the power system domain, different data mining techniques are used to analyse load data (Flath et al., 2012). However, traditional clustering tasks re-cluster all data from scratch whenever an influx of new data arrives.
- The incremental clustering method is essential to solve the problem of grouping growing data for pattern recognition.
- In order to reduce carbon emissions from the production of electricity, the exact consumption patterns of consumer(s) is required.
- It is most essential to keep learning incrementally for ever-increasing consumption data for effective decisions, predictions and problem-solving.
- This can be achieved effectually by applying incremental clustering methods on real ESM data sources.
- Incremental learning can be achieved via incremental clustering easily as well as effectively.

1.1.3 Incremental Clustering Methods

- Clustering is a subcategory of unsupervised machine learning. However, traditional clustering lacks the concept of new learning.
- In Advanced Machine Learning (AML), incremental clustering is a novel concept (Kulkarni and Joshi, 2015). Incremental clustering makes effective use of knowledge for decision-making.
- The knowledge is continuously revised (evolved) as new pieces of information become available over time to stakeholders in power systems. This is called knowledge augmentation (Mulay and Kulkarni, 2013).
- Incremental clustering makes effective use of new information being evolved and the existing knowledge base for accurate decision-making.
- With the influx of new labeled or unlabeled data, incremental clustering either updates the existing clusters or forms a new cluster, as shown in figure 1.3.

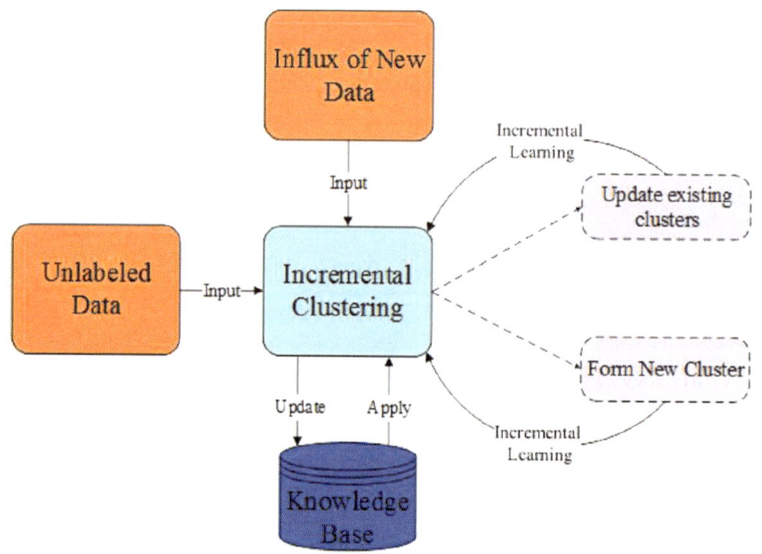

Fig. 1.3. Working Principle of an Incremental Clustering (Source: Chaudhari and Mulay, 2019a)

Problems in Traditional Clustering Approaches

- In the process of decision-making, new data is collected, resulting in new training datasets. This new data can be evolved over a period of time, and is practically challenging for traditional approaches to use.
- The availability of new data may result in new feature vectors that are required to be accommodated selectively.
- In unsupervised clustering, the entire unlabeled data is used in the clustering. Further availability of unlabeled data results in re-clustering, which again becomes a costly venture for traditional clustering.
- The learning does not occur with new data that is being grouped, which could impact the subsequent results in the learning.

The learning takes place in phases and at every stage, some new data is getting available, so the clustering needs to be incremental.

Incremental Clustering vs. Stream Clustering

- Both incremental clustering (Fisher, 1987) and stream clustering (Silva et al., 2013) methods handle continuously changing datasets.
- Stream clustering methods only consider current data and forget historical data. But the old information is equally important as current or new data.
- Incremental clustering methods consider new or current data as well as old data for decision-making. Many applications such as pattern recognition, load profiling, fraud detection and healthcare (Bao et al., 2018) require incremental clustering.

1.2 Incremental Learning Methodology

- Incremental learning involves the effective use of knowledge for decision-making and forecasting/predictions, among others.
- Human beings have the ability to learn and make decisions based on the knowledge they have gained. It is possible for us to relate new scenarios with previous ones. We remember the outcomes and consider the impacts caused in the learning. Thus, we learn, adapt and evolve.

Incremental Learning (IL) is the ability of the learning methodology to make effective use of new information that is being evolved and the existing knowledge base for accurate decision-making (Kulkarni, 2012).

- As shown in Fig. 1.4, with every new learning, the knowledge base is updated and applied further for incremental decision-making.
- IL approaches can be categorized into absolute and selective incremental learning.
- In absolute learning, the learning occurs without consideration of old scenarios, and the new data are analysed separately. Selective learning neither learns from scratch nor retains all feature vectors learned previously (Kulkarni, 2012).

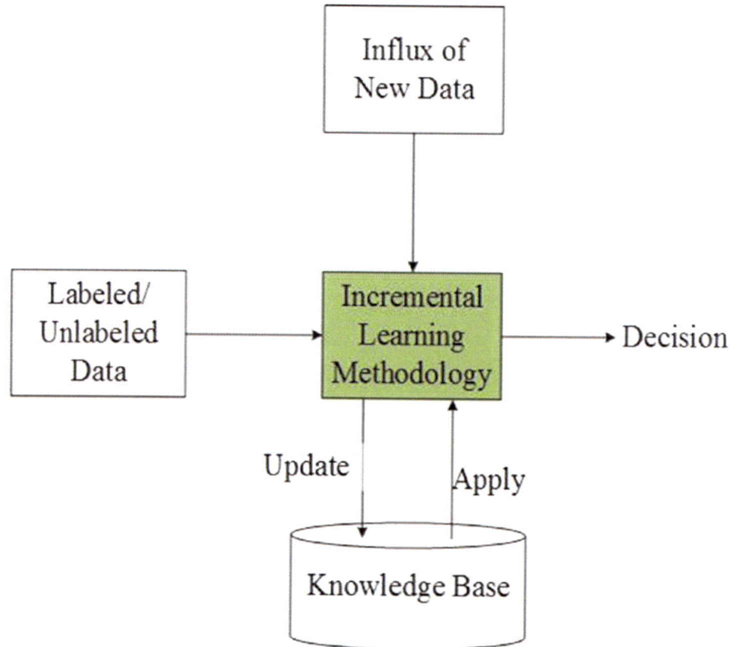

Fig. 1.4. Incremental Learning Methodology Framework Proposed in this Research

1.3 Applications of Electricity Smart Meter Data Analytics

- Electricity Smart Meter Data Analytics (ESMDA) not only considers electricity usage data but also the related socioeconomic data, geographic information, demographic data, etc.
- Therefore, the amount of the data to be analysed will be large and always scaled up over time due to Internet of Things (IoT) and other technologies, and the data will be from different data sources.
- The taxonomy of ESMDA, as shown in figure 1.5.
  The primary applications of ESMDA classified into load analysis, load forecasting, outage management, customer characterization, and so on so forth. The various techniques include time series, dimensionality reduction, clustering, classification, deep learning, etc used by ESMDA to achieve above primary applications (Wang et al., 2018).

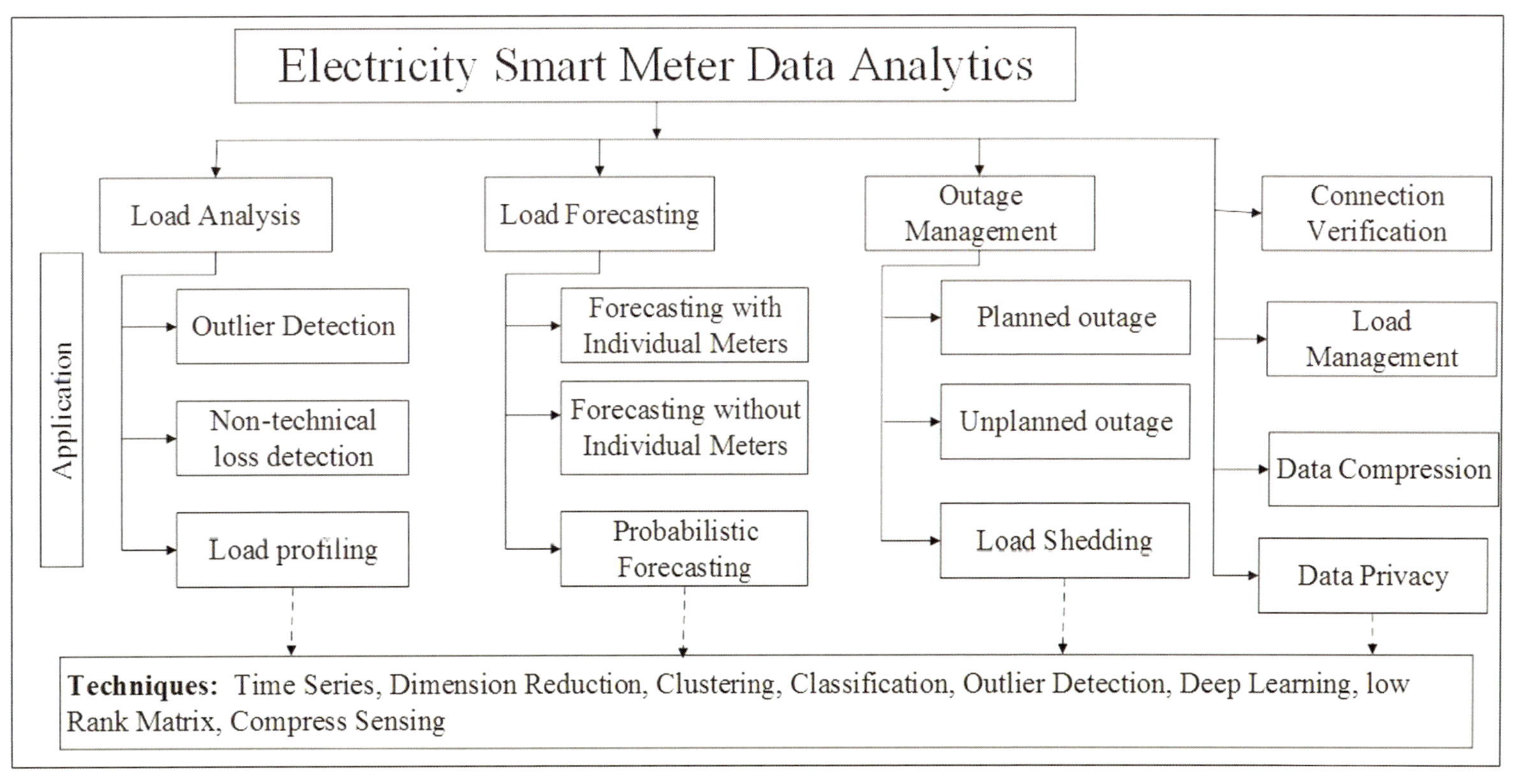

Fig. 1.5. The taxonomy of ESMDA (Source: Wang et al. (2018))

1.3.1 Load Profiling

- The process of mining data for learning a meaningful pattern is called load profiling.
- Load profiling of ESM data is based on geographic regions or time period (Pawar and Momin, 2017). The energy consumption pattern gives information about the time and duration of energy consumed by the consumer.
- A large number of researchers have used clustering techniques to model energy consumption according to different characteristics such as outdoor temperature, calendar attribute, sunset and sunrise time, and socioeconomic factors. For the load profiling framework, K-means is the most commonly used clustering method (Ma et al., 2017).
- Various clustering techniques, such as Self-Organizing Map (Chicco, 2012), Fuzzy c-mean, Hierarchical Clustering and Expectation-Maximization (R. Granell and Wallom, 2015), etc have directly implemented on ESM data. While indirect-clustering has applied to the features of ESM data, that are extracted before clustering as describe in figure 1.6.
- The goal of load profiling is customer segmentation for demand-side management, tariff designing and so on.

Load Profiling

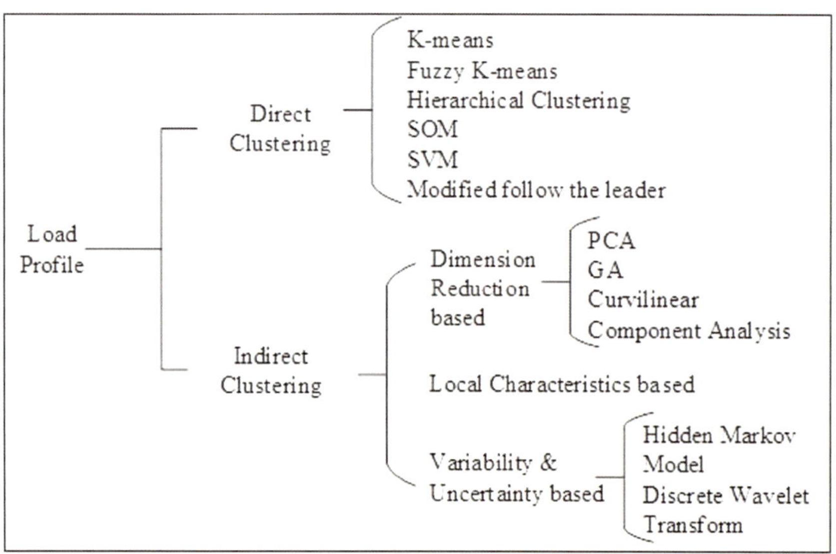

Fig. 1.6. Load Profiling Categories (Source: Wang et al., 2015).

Limitation: Need to learn from scratch when continuously evolving new data

1.3.2 Demand Response Program

- According to the Federal Energy Regulatory Commission (Chatterjee, 1977), Demand Response (DR) is defined as: "Changes in electricity use by end-use customers from their normal consumption patterns in response to changes in the price of electricity over time, or to incentive payments designed to induce lower electricity use at times of high wholesale market prices, or when system reliability is jeopardized."
- DR programs have been developed to modify the energy requirement rather than adjust supply.
- Energy consumption varies due to weather conditions, energy prices, and dynamic behavior of consumers (Chen, 2017). However, it is very complicated to identify the energy consumption patterns of consumers using the conventional system model.
- A paper published by Pabon et al., 2017, shows the use of R programming system to predict the eligibility of a consumer to participate in DR programs by load consumption pattern.

1.3.3 Load Forecasting

- Electricity Load Forecasting is one of the important tasks of electricity generation units and governing bodies.
- Electricity Load Forecasts can be categorized into very short-term, short-term, medium- and long-term forecasting, as the tenure varies from hours, days, and weeks to months. Hence, this classification plays a vital role in decision-making.
- This decision-making is about requirements of building new power plants, allocating funds, catering to increased demands etc. to name a few.
- The most important angle to these forecasts is collaborations between countries or states for fulfilling electricity requirements to provide uninterrupted power supply to individuals and various commercial units of the countries (Fallah et al., 2018).
- Various machine learning techniques used for load forecasting, such as Recurrent Neural Network, Support Vector Machine and Artificial Neural Network, imitate human beings' way of thinking and reasoning to get knowledge from experience and forecast the future load (Kong et al., 2018b).
- The ESMDA intimates energy producers with information on the amount of energy likely to be needed in the future.

1.4 Distributed Clustering Methods

- Clustering algorithms can be applied to a wide range of problems, including wireless sensors networks (Predd et al., 2006), peer-to-peer networks (Ang et al., 2013), power grids (Blaabjerg et al., 2006) and customer segmentation.
- Traditional clustering algorithm requires all data to be located at the central site where they are analysed. As the datasets are geographically distributed across multiple sites and the size of data has grown rapidly, traditional clustering is not suitable for most distributed environments.
- The distributed clustering technique is essential to understand the hidden information present in distributed datasets. This method involves the least amount of communication overhead (Kotary and Nanda, 2020).
- Distributed clustering is expected to perform partial analysis of data at individual sites and then send the partial results as outcomes to a central site where they are aggregated as a global result.

Graphical comparison of centralized and distributed clustering approaches is shown in figures 1.7 and 1.8.

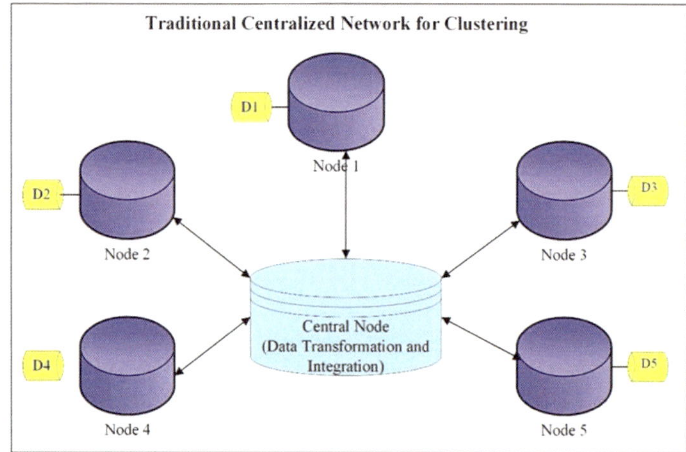

Fig. 1.7. Centralized architecture, where all the data is sent to a central node that performs the clustering task.

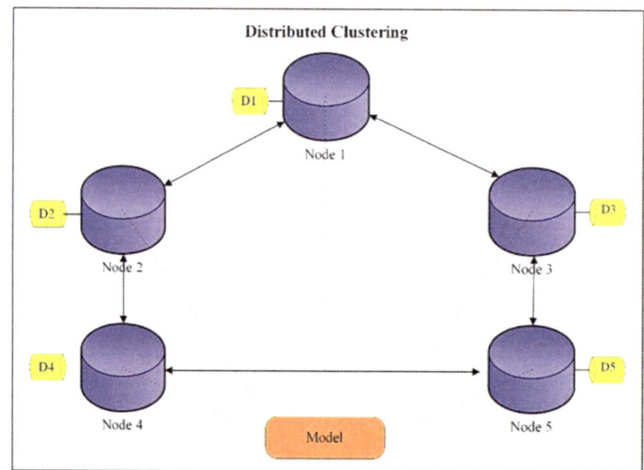

Fig. 1.8. Distributed architecture, where data is distributed over the node, and collaborative learning is performed via the local exchange of information within neighborhoods defined by the communication network.

1.4.1 Distributed Incremental Clustering Methods

- Current distributed clustering strategies customarily deliver overall models by gathering neighborhood results that are gained on each site.
- In Bendechache and Kechadi, 2015; Huang et al., 2017, a new clustering approach is proposed for enormous spatial datasets that are heterogeneous and disseminated. The methodology depends on K-means Algorithm but it generates the range of global clusters progressively. Additionally, this methodology utilizes an elaborated aggregation phase. The aggregation phase is designed so that the general procedure is effective in time and memory distribution.
- In Pfander et al., 2019; Bhimani et al., 2015 proposed a distributed and performance-portable variant of the sparse grid clustering algorithm that is suited for big datasets. The peak performance achieved across all computed kernels and devices was between 43% and 66%.
- In Qin et al., 2017, cooperation among sensor nodes is achieved using a quick and precise appropriated K-means algorithm. In Zhou et al., 2017, data points are described primarily based on a probability density function and a distributed uncertain K-Means clustering is proposed.

1.5 Characteristics of Incremental Learning via Incremental Clustering Related to Electricity Smart Meter Data

1. Learns from the new unlabeled instances of data without discarding previously acquired knowledge: the system presented here considers influx of new data, current ESM data as well as historical data for decision-making.
2. Updates the knowledge base and, thus, the feature vectors with the learning: incremental clustering and cloud computing technologies (Microsoft Azure or alike) and ESM data analytics can improve energy management for both utilities and customers.
3. Formulates the knowledge effectively with any new learning that has occurred: the system presented here first forms basic clusters and then, on an influx of new data, either updates existing clusters or forms a new cluster. Such continuous update of cluster details is useful for knowledge augmentation.
4. Evolves with new scenarios as per the problem requirements: the incremental clustering approach is a way to address dynamically evolving ESM datasets.
5. Learns from new training datasets that would be evolved over a period of time: the system presented here forms clusters based on the electricity consumption of customers and also on the influx of new data. It is further enhanced to form clusters to inform the usage to energy generation units based on hidden patterns, shape of clusters, outage details, theft, etc.

1.6 Datasets for Research

Name	Brief Description	No. of houses	Frequency	Duration	Ref.
ISSDA	Smart meter read data; Pre- and post-trail survey data (Ireland)	6445	Every 30 min	2009/9-2011/1	(ISSDA, 2018)
Low carbon London	Smart meter data; Electricity price data; Appliance and attitude survey data (London)	5567	Every 30 min	2013/1-2013/12	(Schofield et al.2013)
Umass Smart	Residential apartment electricity consumption data (USA)	114 Apartments	Every 1 min	2014/10-2016/12	(Barker et al. 2012)
ENERTALK	Electricity consumption data: aggregate and appliance level (South Korea)	22	15 Hz	2016/11-2017/01	(Shin et al., 2019)
OPSD	Electricity consumption, wind power and solar power production (Germany)	4384	Daily	2006/1-2017/12	(OPSD: Open Power System Data, 2020)
ENTSO-E	Electricity consumption data and weather data (Europe)	9000	Hourly	2014/12-2017/5	(Hirth et al., 2018)
Prayas (Energy Group)	Smart meter consumer data (Pune region)	70	Every 1 min	2018/1-2018/2	(Prayas Group,2018)
IITB	Electricity consumption data from a high-rise residential building inside the IIT Bombay campus (Mumbai)	60	Hourly	2016/12 -2018/6	(Mammen et al., 2018)
I-BLEND	Commercial and residential buildings of an academic institute campus(Delhi, India)	7 buildings	Every 10 min	2013/8-2017/12	(Rashid et al., 2019)

Table 1.1: Summary of Several Open Load Datasets

Datasets for Research (International)

- Customer Behavior Trials (CBTs): The commission for energy regulation propelled a smart metering project in Ireland to lead CBTs. The objective of the Irish Social Science Data Archive project is to determine if smart meter can shape energy utilization practices over an assortment of socioeconomics and dwelling sizes (ISSDA, 2018).
- Low Carbon London (LCL): The objective of the LCL project is to identify the impact of low carbon technologies on London's electricity distribution network (Schofield et al., 2013). It includes smart meter data as well as tariff data.
- Umass Smart: The goal of this dataset is to optimize home energy consumption. It includes electrical consumption and generation as well as environmental datasets (Barker et al., 2012).
- ENERTALK: This South Korean dataset includes aggregate and per-appliance residential electricity measurements of 22 houses (Shin et al., 2019).
- OPSD: Open Power System Data is a Germany-wide total of electricity consumption, wind power production and solar power production for 2006-2017. Electricity production and consumption are reported as daily totals in gigawatt-hours (GWh) (OPSD, 2020).
- ENTSO-E: European Network of Transmission System Operators for Electricity dataset granularity is half-hourly (Hirth et al.,2018). ESM data reflect the total electricity consumption in this country. (from 2014/12 to 2017/5)

Datasets for Research (India)

- Prayas Energy Group: This dataset is supported by the Prayas Energy Group, Pune, India. The dataset contains minute-level electricity consumption data from 77 houses (Prayas Energy Group, 2018).
- IITB-Smart Energy Informatics: This IIT Bombay campus dataset consists of electricity consumption of a residential apartment (3BHK building) (Mammen et al., 2018).
- I-BLEND: This Indian Buildings Energy Consumption Dataset includes commercial and residential blocks (Rashid et al., 2019) with a 10-minute sampling rate.

LCL, CBT, and Umass datasets are often used in the existing literature. In addition, IITB, ENERTALK, and I-BLEND are newly released datasets.

Table 1.1 summarizes the publicly available ESM datasets at national and international levels.

1.7 Problem Analysis

- Current solutions for load profiling do not consider the incremental learning via incremental clustering algorithm. It is useful to incorporate an incremental clustering method for reducing energy consumption.
- Incremental learning is important for the ever-growing electricity smart meter data. Incremental learning via incremental clustering should be deployed to utilize updated and clustered smart meter data as knowledge for further mining on various distributed platforms.
- Outliers can be detected from smart meter data for better understanding of abnormal consumption of energy and analysis of electricity smart meter data.
- As data increases multifold and needs to be accommodated, the problem of re-clustering is most evident when new data is made available.
- Knowledge augmentation via incremental clustering for smart meter data analytics can improve energy management for both consumers and utilities.
- The datasets available for the research based on either electricity smart meter or any smart meter can detect the hidden and incremental patterns of load shedding area wise to inform customers in advance.
- To improve the performance and efficiency of load profiling, the features determined by principle component analysis in the post-clustering phase need to be validated.
- The Expectation-Maximization (EM) Algorithm (Li and Nehorai, 2018) is incremental clustering algorithms, and working in two different phases with dependency on user inputs.

1.8 Summary of Chapter 1

- Conventional clustering algorithms require the end user to input various details.
- The inputs required from the end users may be the number of clusters to be formed, distance measure to be used, the assumption of centroids, etc. in addition to inputting raw datasets for forming the clusters.
- To enter all these details as input to the clustering algorithm, the end user needs to be completely aware of the nature of the raw dataset.
- If the user is not aware of the dataset, wrong details may be entered to the clustering algorithm, thereby hampering the quality of clusters.
- Therefore, there is a need for a system wherein the dependency on the end user is completely removed.
- The next chapter covers a Problem Identification carried out for the development of ESMDA.

Chapter 2

Problem Identification

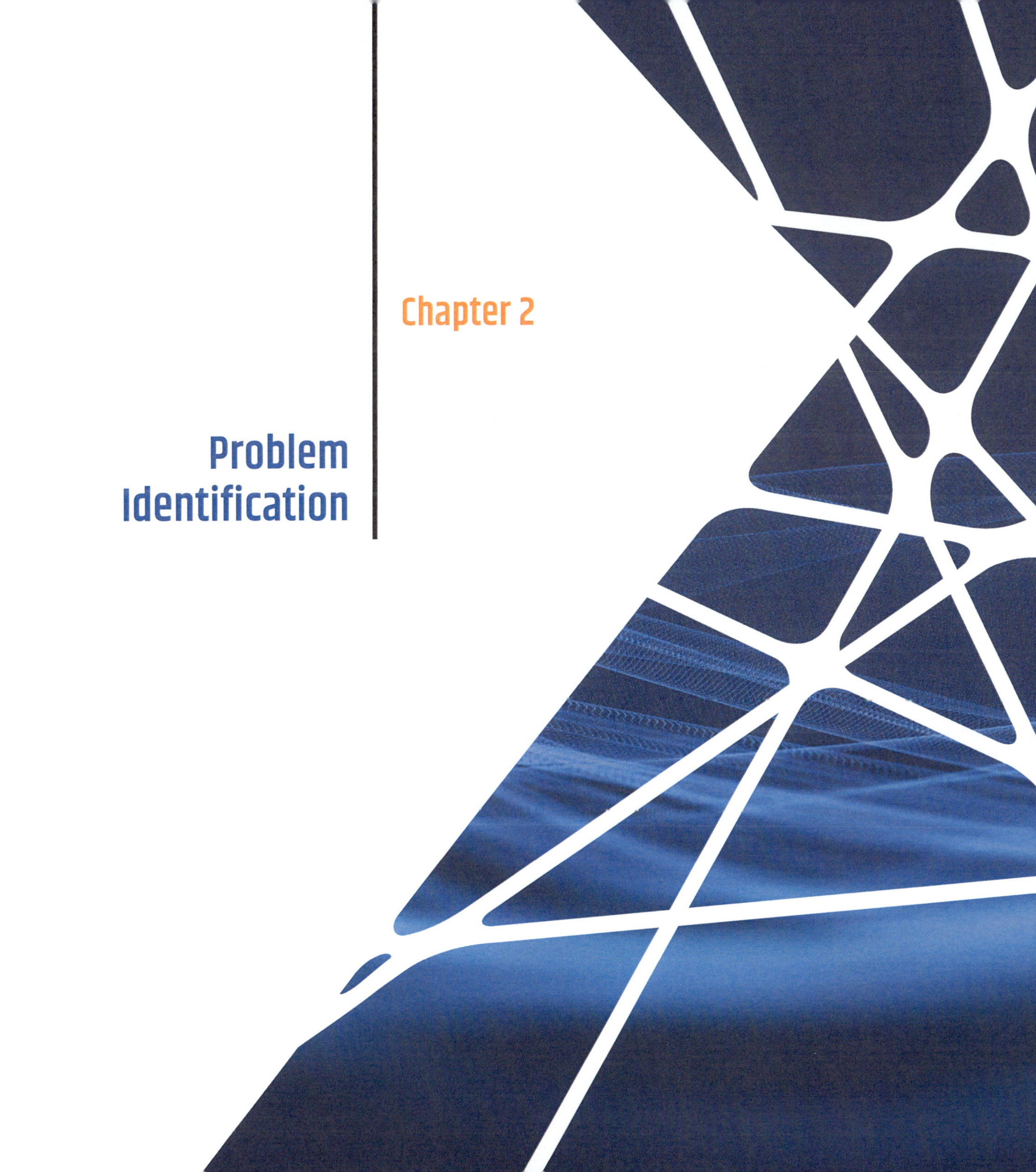

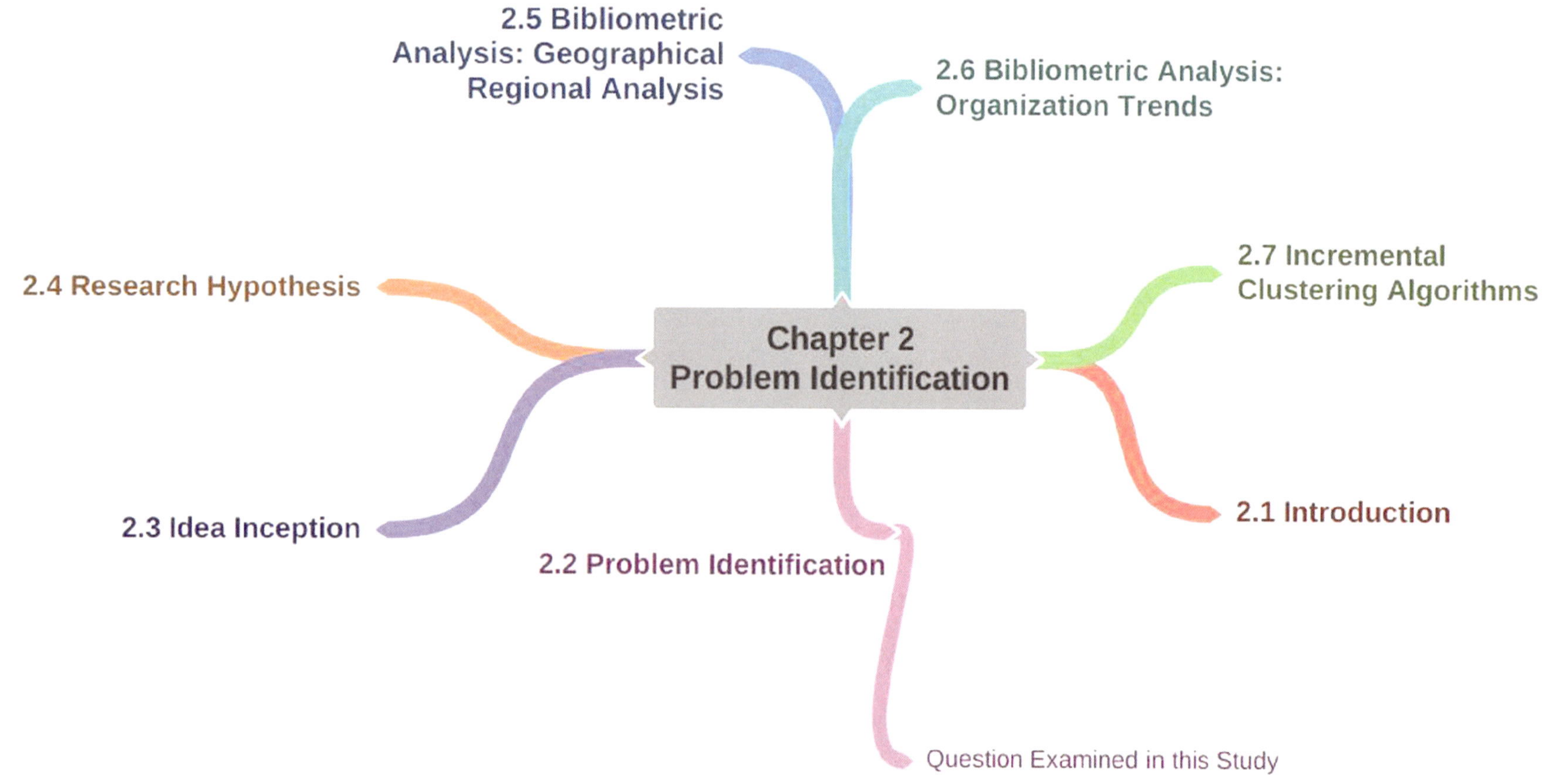

Mind map for chapter 2

Source: https://coggle.it/

2.1 Introduction

- To achieve worldwide data analysis related to ESM data and to achieve broader perspectives, it is essential to deploy incremental clustering algorithms on distributed platforms, which will enable them to accept data from varied sources, analyse it and produce distributed worldwide solutions.
- Given the volume of ESM data and the number of data types involved, an incremental clustering method is highly complex.
- Microsoft Azure and similar platforms provide the data security, privacy, networking, and processing power necessary to handle incremental clustering analytics.

2.2 Problem Identification

- The research aim is to develop a distributed, robust and efficient "incremental clustering method for data analysis" using a Microsoft Azure environment or the like. The proposed method is designed to handle incremental data generated using ESM.
- The system presented here is able to mine the hidden pattern of energy consumption such as load composition, area-specific learning, season-wise learning, time (day/night) - specific learning, etc.
- This system could be used for targeted demand-response programs and services such as targeted marketing and promotions based on household types, as it is the outcome of the research of many years and distributed in nature with iterative data handling capabilities.

2.2.1 Questions Examined in this System

The possible clarification of questions pertaining to this system, as mentioned below:

1. How to process smart meter data in a real-time and efficient manner?
2. How to incorporate the concept of a distributed system to improve the performance and efficiency of the existing algorithm on time series data?
3. How to formulate the knowledge effectively with any new learning that has occurred?
4. How to effectively visualize data analysis results?
5. Who will be the beneficiaries?

2.3 Idea Inception

The proposed system is intended to develop an incremental clustering algorithm for ESM data analysis, which will cover the following objectives:

1. To redesign an Expectation-Maximization (EM) algorithm that can accommodate a large dataset incrementally, because the EM algorithm heavily depends on the initial guess and can be slow, especially when the number of samples are large
2. To implement the distributed system (Cloud4CGMIC) on Microsoft Azure or a similar platform
3. To develop load profiling using incremental learning via incremental clustering. to detect outliers
4. To perform a comparative analysis of the proposed and predefined system using various smart meter datasets

2.4 Research Hypothesis

The research hypothesis is a tentative assumption made in order to draw out and test its logical or empirical consequences. The hypothesis follows from the stated research questions are:

- The analytic results will be used to improve demand response and energy forecasting
- The proposed analytics platform is adaptable enough to be used in other areas in which significant amounts of data is generated
- Better load profiles and in turn systematic life cycle of energy from generation to utilization without or with minimum losses.

2.5 Bibliometric Analysis: Geographic Regional Analysis

Fig. 2.1. Countries Involved in Publications related to Electricity Smart Meter Data Analysis using Incremental Clustering Algorithm ESMDAIC (accessed on 19th Nov 2019)
Source: https://www.imapbuilder.net/

Figure 2.1 is drawn using https://www.imapbuilder.net/ showing geographical regional location attentiveness of published papers in the area of ESMDAIC from WoS and Scopus. It is visible from the map that European countries are the prominent publishing countries for undertaken study.

2.6 Bibliometric Analysis: Organization Trends

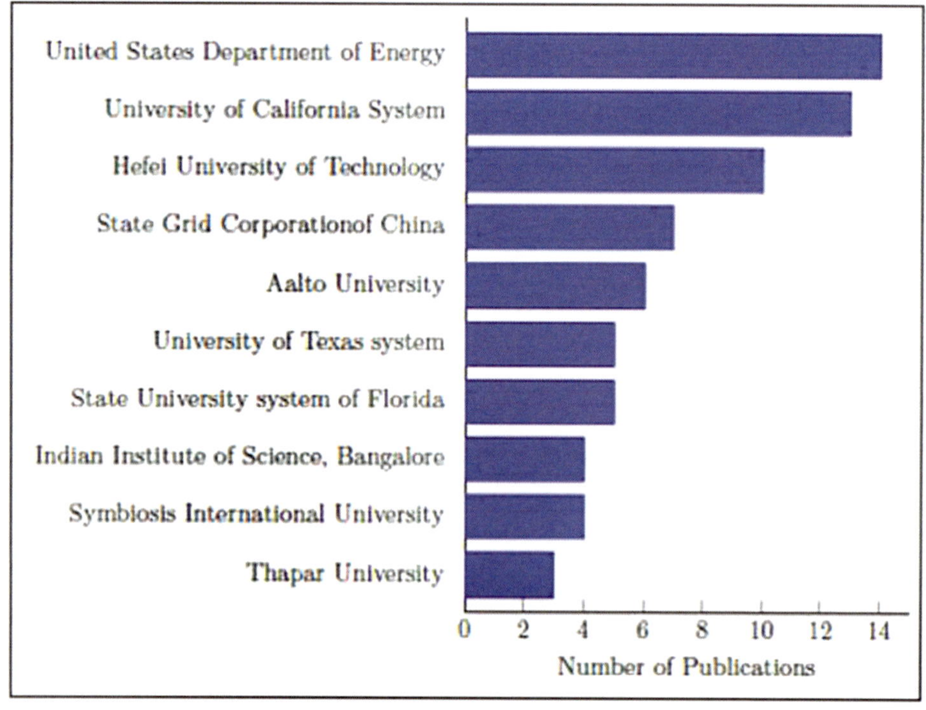

Fig. 2.2. Top 10 Most Popular Affiliations (accessed Scopus.com on 19th Nov 2019) *https://www.scopus.com/*

Figure 2.2 shows the number of documents affiliated with the comparative universities, and the more affiliation was done from the United State Department of Energy, followed by the Symbiosis International University. It shows that mostly affiliation is done at California Universities after that at Chinese University.

2.7 Incremental Clustering Algorithms

- In the literature, various unsupervised machine learning techniques, specifically clustering analysis, are commonly used to categorize the pattern of the load, analyse residential electricity consumption data and extract consumption patterns (Chaudhari and Mulay, 2019a).
- The existing methods that are incremental in nature are DBSCAN (Ester et al., 1998), BIRCH (Zhang et al., 1997), COBWEB Fisher (1987) and EM (Moon, 1996).
- In incremental K-means (IK-means), clustering algorithm is applied to a dynamic database where the data may be frequently updated. This approach measures the new cluster centers by directly computing the new data from the means of the existing clusters instead of rerunning the K-means algorithm. IK-means produces k number of clusters effectively (S. Chakraborty , 2011).
- In Vasighi and Amini, 2017, a novel growing self-organizing map is proposed, which highlights steady bunching and a dynamic system structure. However, loads of the model are hard to relegate at first, prompting an inadequate adjustment.
- In Wang et al., 2019, performance is improved using information with an imbalanced circulation; the Incremental Local distribution-based clustering method with Bayesian Adaptive Resonance Theory (ILBART) is proposed.
- All these current algorithms require prior knowledge from the user, such as the number of clusters, radius, number of minimum points, distance measures, etc. If the end user selects the wrong input(s), it hampers the quality of the clusters.
- These issues are addressed (Mulay and Kulkarni, 2013; Kulkarni and Mulay, 2013) using a parameter-free three-phase algorithm called Closeness Factor-Based Algorithm (CFBA).
- CFBA is a one-of-its-kind data clustering method, which accommodates the arrival of new data effectually and automatically. It is further extended to Threshold-Based Clustering Algorithm (TBCA) to analyse diabetic patients' clinical parameters (Mulay, 2016).
- TBCA is a probability-based algorithm; hence, there is a restriction on the range of the threshold value from 0 to 1.
- To further enhance the threshold range from -1 to +1, a new variant of CFBA has taken shape called Correlation-Based Incremental Clustering Algorithm (CBICA) (Mulay and Shinde, 2017; Shinde and Mulay, 2017).
- Another variant of CFBA is Log Likelihood-based Gradational Clustering Algorithm (L2GA). L2GA uses log likelihood as a similarity measure. (Chaudhari, A. and Mulay, P. (2019b)).

Chapter 3

Electricity Smart Meter Data Analysis using Incremental Clustering Algorithm (ESMDAIC)

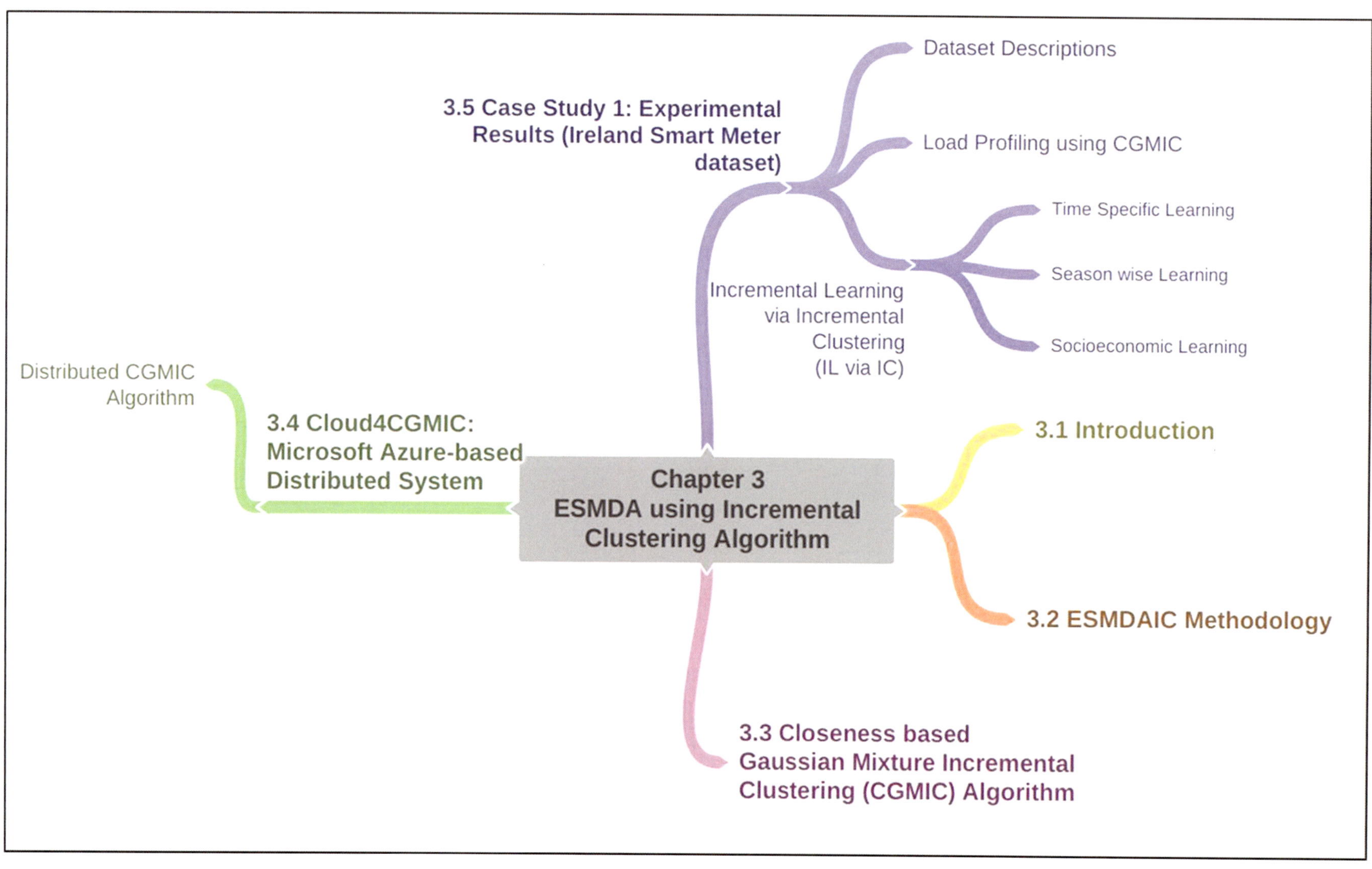

Mind map of part 1 of Chapter 3 showing details about sections 3.1 to 3.5

Source: https://coggle.it/

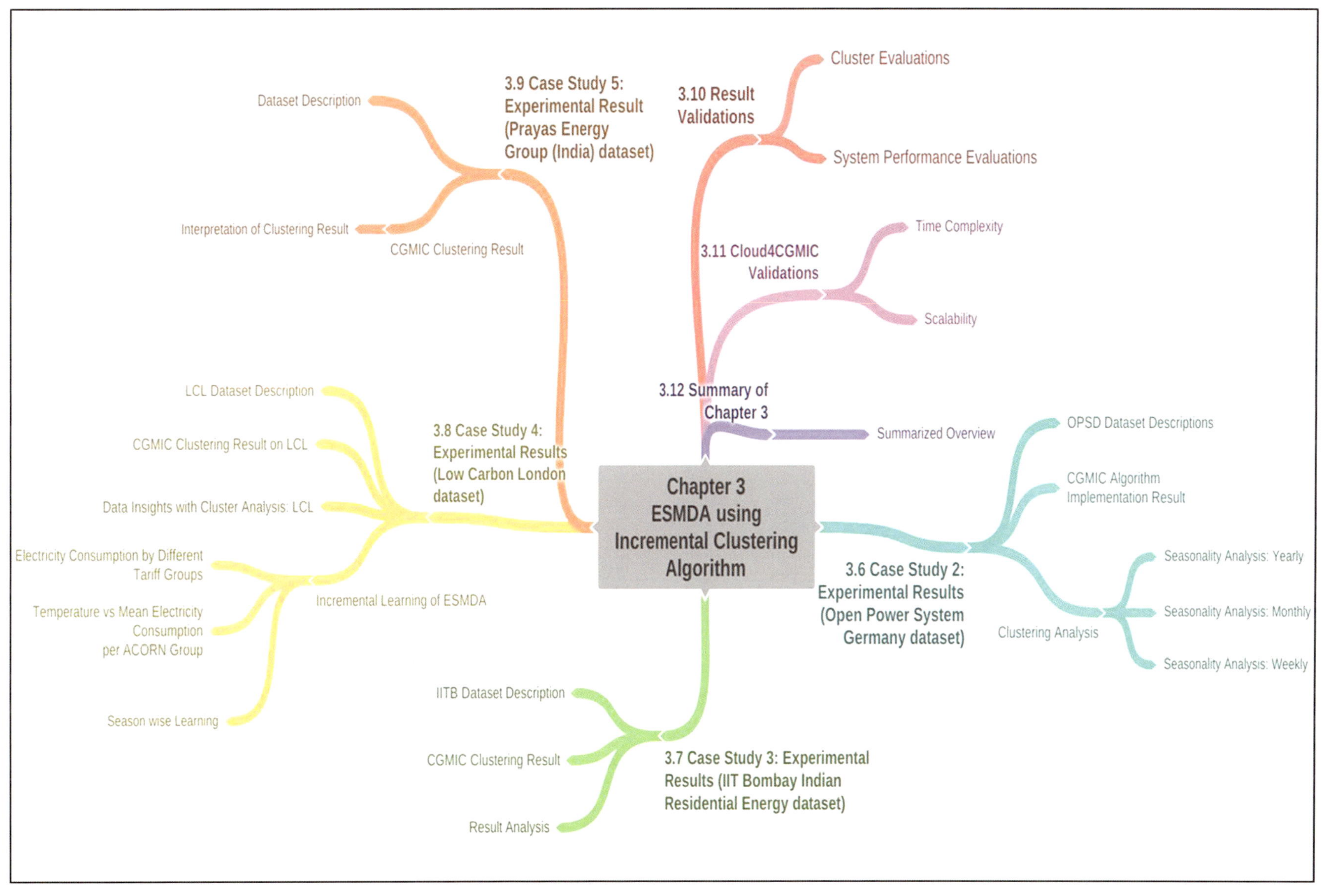

Mind map of part 2 of Chapter 3 showing details about sections 3.6 to 3.12

Source: https://coggle.it/

3.1 Introduction

- The role of Incremental Clustering (IC) in the ESM environment is described in this chapter.
- This research is about the development of the IC algorithm for analysis of ESM data.
- The Closeness-based Gaussian Mixture Incremental Clustering (CGMIC) algorithm is designed to improve overall electricity energy management holistically, enabling reduced electricity consumption and carbon emission much faster through distributed computations on many machines.
- IC is a dynamic approach of grouping related data series together. IC algorithms start with the creation of basic clusters with initial datasets available.
- Once the basic clusters are ready, on the arrival of new data, the IC algorithm effectually accommodates the new data by either appending the existing clusters or forming new cluster(s) automatically.
- When the basic clusters are made and analysed, patterns are generated. On arrival of new data every time, these clusters may observe changes in their individual structures, clusters representatives, and threshold range.
- Hence, there is a need for an incremental learning algorithm that captures all these details and learns from every iteration of new data.
- The EM algorithm can be slow, especially when the number of samples is large. The local maximum likelihood solution for EM depends heavily on the initial guess of a number of Gaussian components in the mixture model (Li and Nehorai, 2018).
- Choosing the correct attributes (initial guess) is potentially the most important aspect of a successful clustering.
- The present study proposes a Closeness-based Gaussian Mixture Incremental Clustering (CGMIC) algorithm that combines the Closeness Factor-Based Algorithm (CFBA) (Mulay and Kulkarni, 2013) with the EM algorithm.
- The informed initial guess is often close to a maximum likelihood estimate and thus needs fewer EM iterations, which makes the EM algorithm fast enough for incremental datasets.
- Also, the quality of the informed initial guess potentially improves the accuracy of the EM output.
- Given the volume of data and the number of data types involved, ESM data analytics using CGMIC algorithm are highly complex.
- Microsoft Azure or similar platforms provide the data security, privacy, networking, and processing power necessary to handle incremental clustering analysis.

3.2 ESMDAIC Methodology

The development tasks of the algorithm divide the whole process into the different phases, also known as the milestones of this research, as shown in figure 3.1.

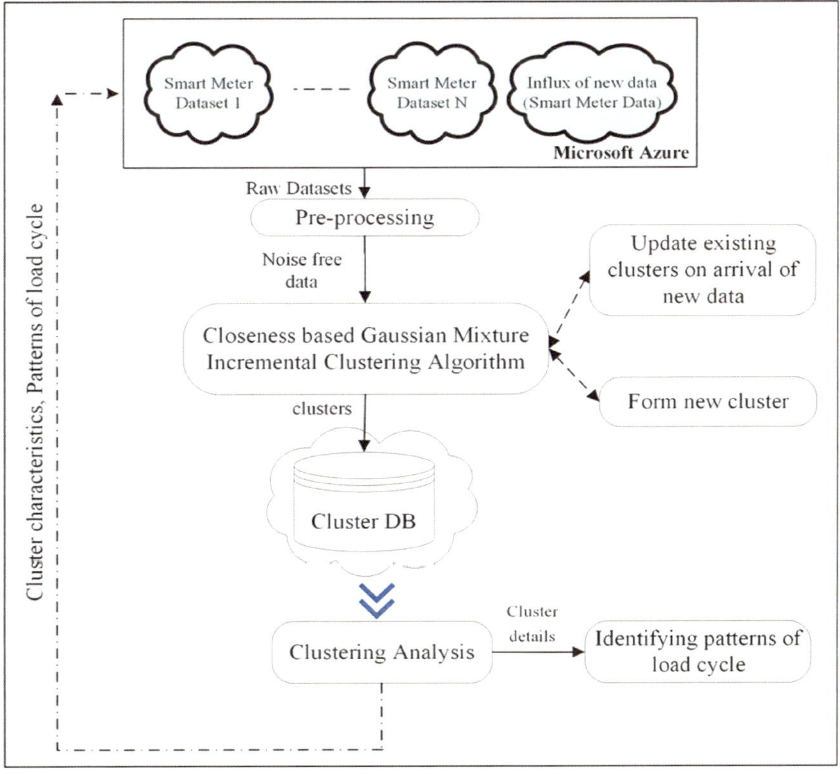

Fig. 3.1. Logical Flow of ESMDAIC System (Source: Chaudhari and Mulay, 2020a)

ESMDAIC Methodology

1. **Data collection:** Data acquisition of existing ESM datasets spanning across Ireland (ISSDA, 2018)
2. **Pre-Processing:** In order to increase the detection accuracy, data is cleansed using data pre-processing techniques.
3. **Incremental Clustering:** The Closeness-based Gaussian Mixture Incremental Clustering (CGMIC) Algorithm accommodates the influx of new data seamlessly for learning, prediction, decision-making, etc. Different values are computed to update the existing clusters or to form new clusters.
4. **Post Clustering:** To test and validate the system using the following techniques:
   - **Cluster Computing-Related Evaluation:** It is used to evaluate the goodness or purity of the clustering result. There are two types of cluster computing evaluation used in this research:
     - Internal Measures: Compactness, Separation, Dunn index, Davies-Bouldin Index
     - External Measures: Rand Index, Normalized Mutual Information, F-indicator
   - **Incremental Clustering-Related Evaluation:** The data is given in a varied, arbitrary order. This test confirms order independence and bchavior of the proposed algorithm in terms of cluster purity, which is missing in other professional algorithms.
5. **Analysis:** Comparative analysis is performed using different clustering algorithms such as EM and DBSCAN.

6. **Intelligent Learning:** There are two types of learning from the proposed system:
    - Direct Learning: Ranking the cluster according to the frequency of updates. Over the iteration, if any one of the clusters shows few or no updates, that cluster is called as an outlier (we introduce). An existing algorithm like DBSCAN finds the outlier by calculating the distance between points (the Euclidean distance, for example) and look for points which are far away from others.
    - Indirect Learning: Reduction of energy (electricity) consumption.

3.3 Closeness-based Gaussian Mixture Incremental Clustering (CGMIC) Algorithm

- The CGMIC Algorithm is an enabler for electricity management, empowering consumers to save and manage their electricity consumption. It also effectively handles incremental data and mines hidden information. This advanced data analysis extract requires knowledge from a large amount of ESM data.
- The novel and inventive features of the proposed algorithm are as follows:
    - Parameter-free, i.e., there is no need to select the number of cluster initialization (like CFBA)
    - Cluster formation first, less complex to implement, and converge guaranteed
    - Cluster ranking during iterations for outlier detection
    - Learns from the influx of new data, without discarding previously acquired knowledge
    - Incremental learning achieved for automatically suggesting groups of clients for specific actions, such as commercial offers for energy reduction.

Cloud4CGMIC Algorithm steps

Input: $I_z = \{I_{z1}, I_{z2}, \ldots, I_{zn}\}$ a set of n d- dimensional time series smart meter raw datasets; M_{itr}: a maximum number of iterations, converge criteria ϵ for loglikelihood.

Output: A series of the cluster stored in clusterdb

Phase I: Formation of Basic Clusters

while *change in loglikelihood (llh) is greater than* ϵ *and* M_{itr} *has not been reached* **do**

  for $i = 1$ *to* n **do**

    i) Consider every two time series I_{z1} and I_{z2}. $I_{zn}(l)$ is the point l in series n. Sum(l) is the total of the corresponding parameters of the series considered.

    ii) The Relationship Probability(RP) of I_{z1} is calculated as ratio of first series to the sum of the corresponding parameters

    iii) Closeness(CN)(Mulay and Kulkarni, 2013) between series are

$$CN = \frac{\sum_{l=1}^{n}((Er(l)^2 * \sqrt{(Sum(l))}}{\sum_{l=1}^{n}\sqrt{Sum(l)}}$$

    where $Er(l) = \dfrac{RP * Sum(l) - I_{zn}(l)}{\sqrt{Sum(l) * RP * (1 - RP)}}$

    iv) number of cluster (k), Mean (μ), Variance (Σ) are stored in clusterdb

  Initialize: Set μ, Σ k, prior probability (Π) from clusterdb and llh=- ∞

  for $i = 1$ *to* n **do**

    for $j = 1$ *to* k **do**

      Posterior Relationship Probability (Fiez and Ratliff, 2019)

$$PP(\frac{C_i}{I_{zi}}) = \frac{PP(C_j)PP(I_{zj}/C_i)}{\sum_{i'} PP(C_i')PP(I_{zi'}/C_i')}$$

      Prior Relationship Probability

$$\text{PRP}(I_{zi} \| \Pi, \mu, \Sigma) = \Pi_j * PP(I_{zi}|\mu_j, \Sigma_j)$$

$$\mu_i = \frac{\sum_{j=1}^{n} I_{zj} * PRP(C_i/I_{zj})}{\sum_{j=1}^{n} PRP(C_i/I_{zj})}$$

$$\Sigma_i = \frac{\sum_{j=1}^{n}(I_{zj} - \mu_i)^2 * PRP(C_i/I_{zj})}{\sum_{j=1}^{n} PRP(C_i/I_{zj})}$$

$$llh = llh + \log(PRP(C_i)PRP(I_{Xi}/C_i))$$

Phase II: On the influx of new data either updation of the existing cluster(s) or formation of the new cluster(s)

3.4 Cloud4CGMIC: Microsoft Azure-based Distributed System

- This research used Microsoft's Infrastructure as a Service.
- The aim is to use the Azure services for faster processing and handling the complexity of dynamically growing data analysis.
- Microsoft Azure provides the processing power to distribute the CGMIC and handle the dynamically growing data.
- Cloud4CGMIC runs on the Microsoft Azure Cloud platform to process distributed ESM data, accommodates the influx of new data and forms clusters dynamically as shown in figure 3.2.
- Internally, an application is composed of one or more web and/ or worker roles, residing in and executing the application logic on Virtual Machines (VMs) (Mrozek et al. (2015)).
- Web role VMs and worker role VMs are Windows Servers with and without IIS installed, respectively.
- The proposed Cloud4CGMIC processing model shown in figure 3.3. The web role provides a system front-end for users, manager role does the distribution of the closeness value calculation process, and the worker roles execute the CGMIC algorithm and predict the electricity consumption patterns.
- The control instructions and parameters within VMs are transferred through respective queues.

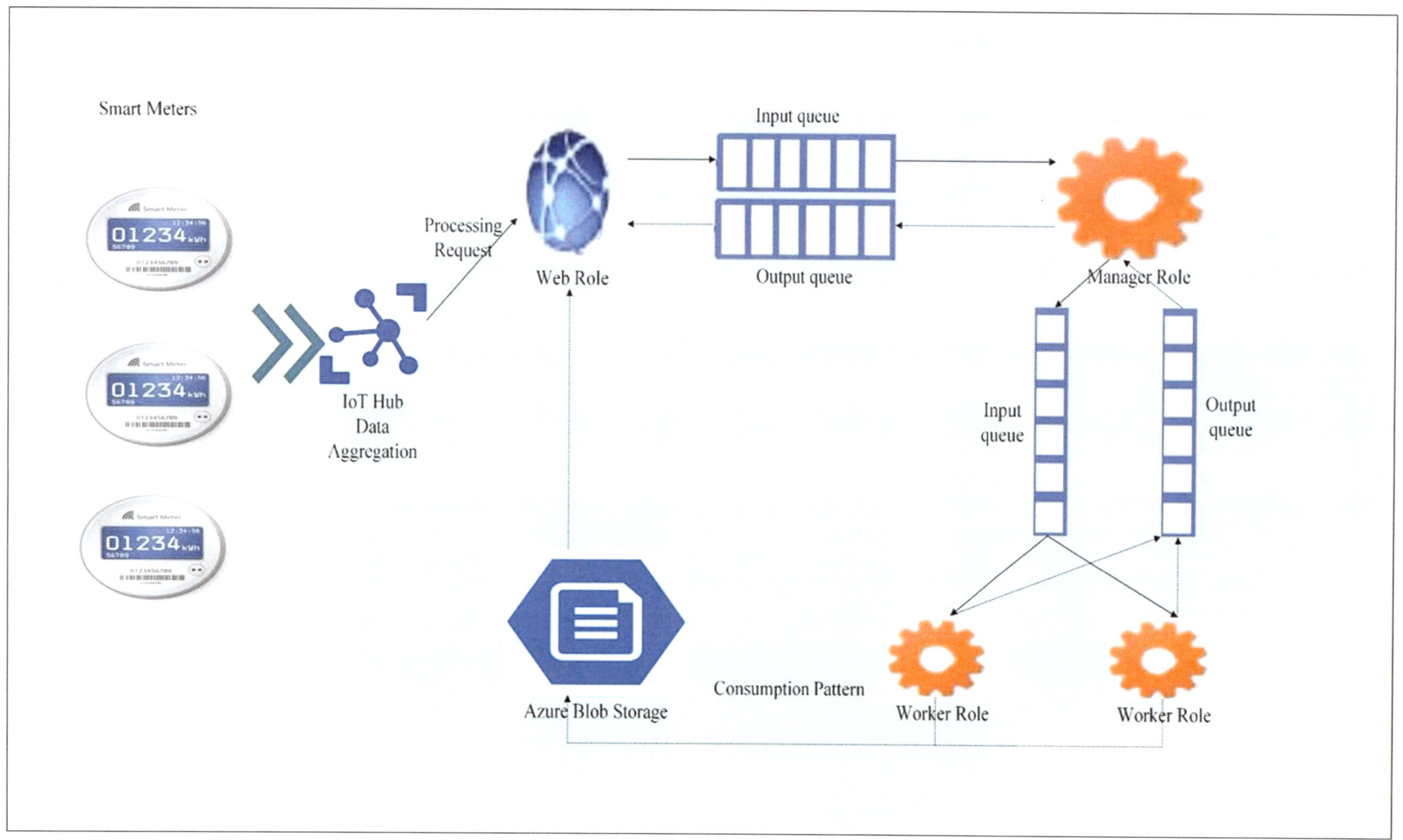

Fig. 3.2. Cloud4CGMIC Architecture(Source: Chaudhari and Mulay, 2020b)

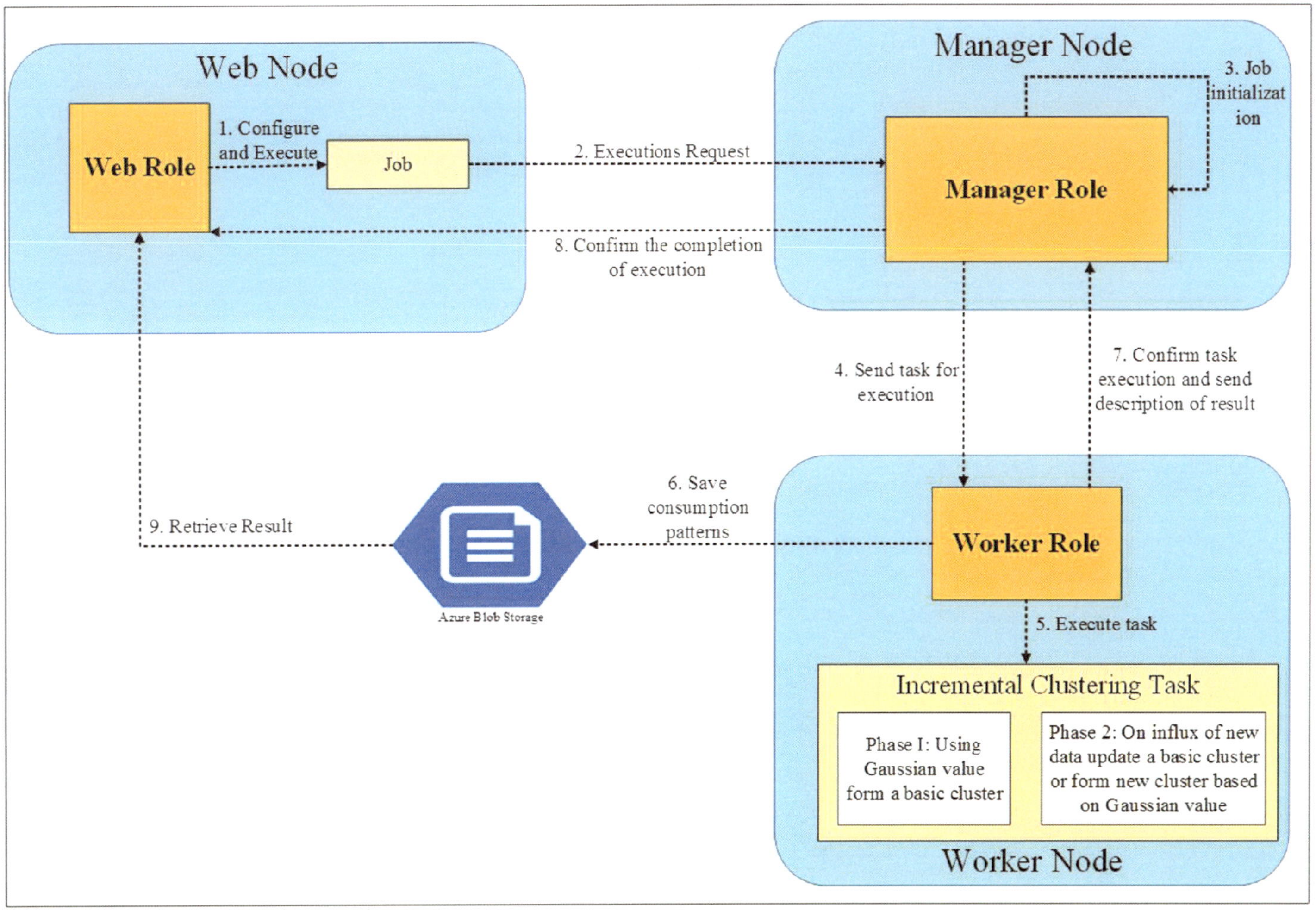

Fig. 3.3. Cloud4CGMIC Processing Model (Source: Chaudhari and Mulay, 2020b) 

3.4.1 Distributed CGMIC (DCGMIC) Algorithm Steps

1. Collection of distributed ESM data using IoT Hub service of Microsoft Azure
2. Submission of the dataset to the web role for processing request; every request causes the creation of a prediction job
3. These datasets are serialized to the message by using Azure services and then consumed by successive worker role instances
4. Worker role instances process and execute CGMIC algorithm by using the virtual machines, stream analytics, and machine learning services of Microsoft Azure
5. Finally, the result is saved in terms of clusters using Azure file storage service
6. On arrival of new data, either existing cluster(s) are updated or new cluster(s) are formed
7. Worker role confirms the completion of the task to the web role through output queue
8. Electricity consumption patterns are retrieved from the Azure storage.

3.5 Case Study 1: Experimental Results (Ireland Smart Meter Dataset)

The dataset used in this research is from Ireland Smart Meter (ISSDA, 2018) data.

3.5.1 Dataset Description

- The dataset used in this research concerns the electricity consumption of 6,445 buildings recorded using ESM installed by the Irish Commission for Energy Regulation (CER), Ireland (ISSDA, 2018) from 2009 to 2010.
- CER performed ESM customer behavior trials for identifying the pattern of residential electricity consumption.
- ESM has been distributed across the spatial locations of Ireland residents.
- The dataset set contains two types of information: electricity consumption data and related socioeconomic data, including the weather conditions at meter installation sites, customer data and geographic information.
- The ESM dataset comprised 4,232 meters worth of data of residential customers with a granularity of 30 min (48 readings per day).
- After pre-processing (removal of noise and missing values), it provided 2,995 meters worth of data. Total electricity consumption dataset contains more than 6 million instances.

Ireland Smart Meter Dataset

Figure 3.4 depicts the overall electricity consumption pattern of residential customers over one year. The average consumption graph shows that day begins at 12:00 am and the peak is at 8:00 am. Majority of the electricity consumption occurs in the morning between 7:00 am to 9:00 am and in the evening between 6:30 pm and 10:00 pm.

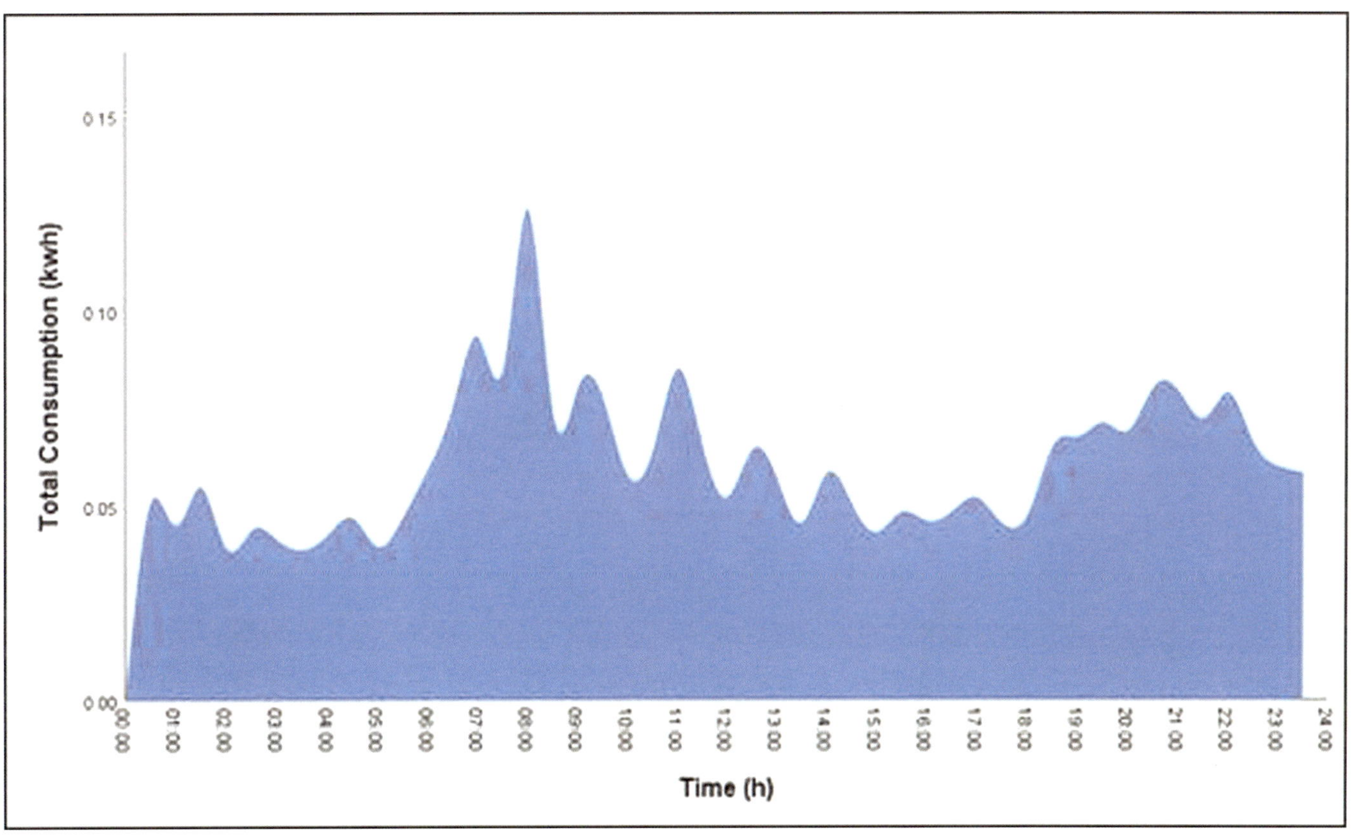

Fig. 3.4. Average Electricity Consumption of Residential Sectors for the year 2010

3.5.2 Load Profiling using CGMIC

- The DCGMIC system is used to understand the consumption patterns of ESM data.
- The CGMIC algorithm uses closeness value and Gaussian mixture to automatically decide cluster members, and the number of clusters is determined using probability and error-based computing.
- Figure 3.5 depicts the clusters obtained from the CGMIC algorithm applied on Ireland's dataset. Figure 3.5(a) presents phase one implementation of the CGMIC algorithm, which results in three basic clusters. The influx of new data fits into the existing cluster as seen in figure 3.5(b) (given in orange). We observe from figure 3.5(c) and 3.5(d) that following an influx of new data, some of the data do not fit into existing clusters because the closeness value is different, so the CGMIC algorithm automatically formed new clusters number four and five, respectively.
- CGMIC divides 2,995 ESM into five clusters. Consumers are grouped based on electricity usage patterns/behaviors.

Fig. 3.5. CGMIC Clustering Results on Ireland Smart Meter dataset: (a) describes the basic clusters; (b) shows update in existing clusters; (c) and (d) show formation of new clusters. (Source: Chaudhari and Mulay, 2019b)

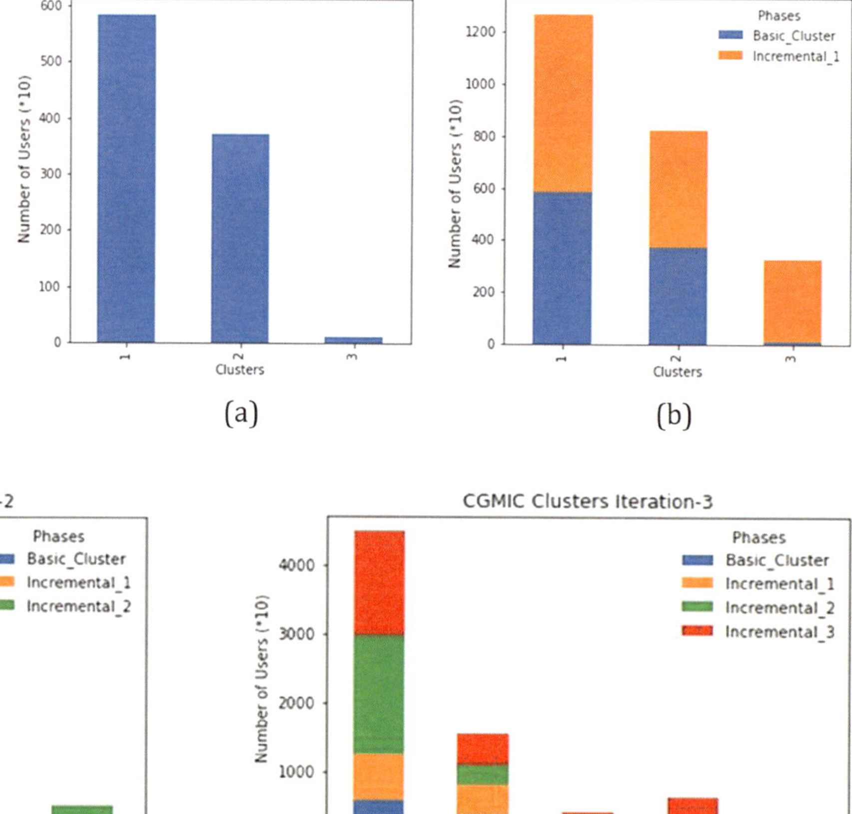

- The clustering result (cluster 1 to 5) and details of the patterns of consumption are summarized in figure 3.6, which shows the overall cluster-wise consumption patterns. Cluster 3 contains the consumers who have consumed the highest electricity, and cluster 5 consists of low demand consumers who consume stable and very less electricity.

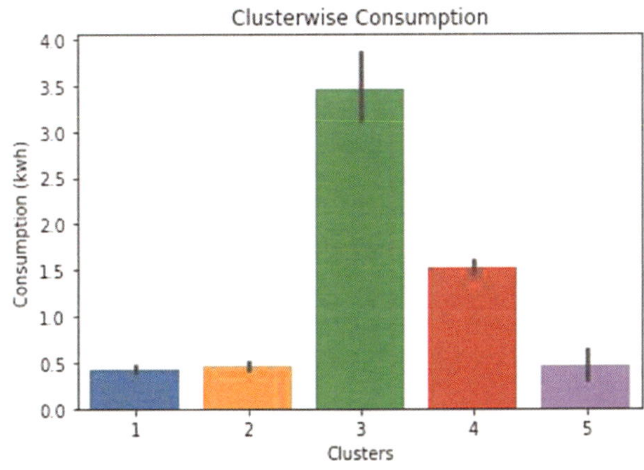

Fig. 3.6. Overall Cluster-wise Consumption Pattern 

3.5.3 Incremental Learning via Incremental Clustering (IL via IC)

- The IL via IC is the ability to make effective use of the new information that is evolved and the existing knowledge base for accurate decision-making.
- Incremental Learning via Incremental Clustering (IL via IC) does the following learning:
    - Time-specific Learning
    - Season-wise Learning
    - Socioeconomic Learning

Time-specific Learning

- Figure 3.7 depicts time-specific learning from clusters which are updated incrementally in various iterations of CGMIC algorithm on Ireland dataset. Consumers are grouped based on electricity usage patterns/behaviors.
- These five clusters help consumers to understand their own electricity consumption patterns and shift their load from on-peak hours to off-peak hours.

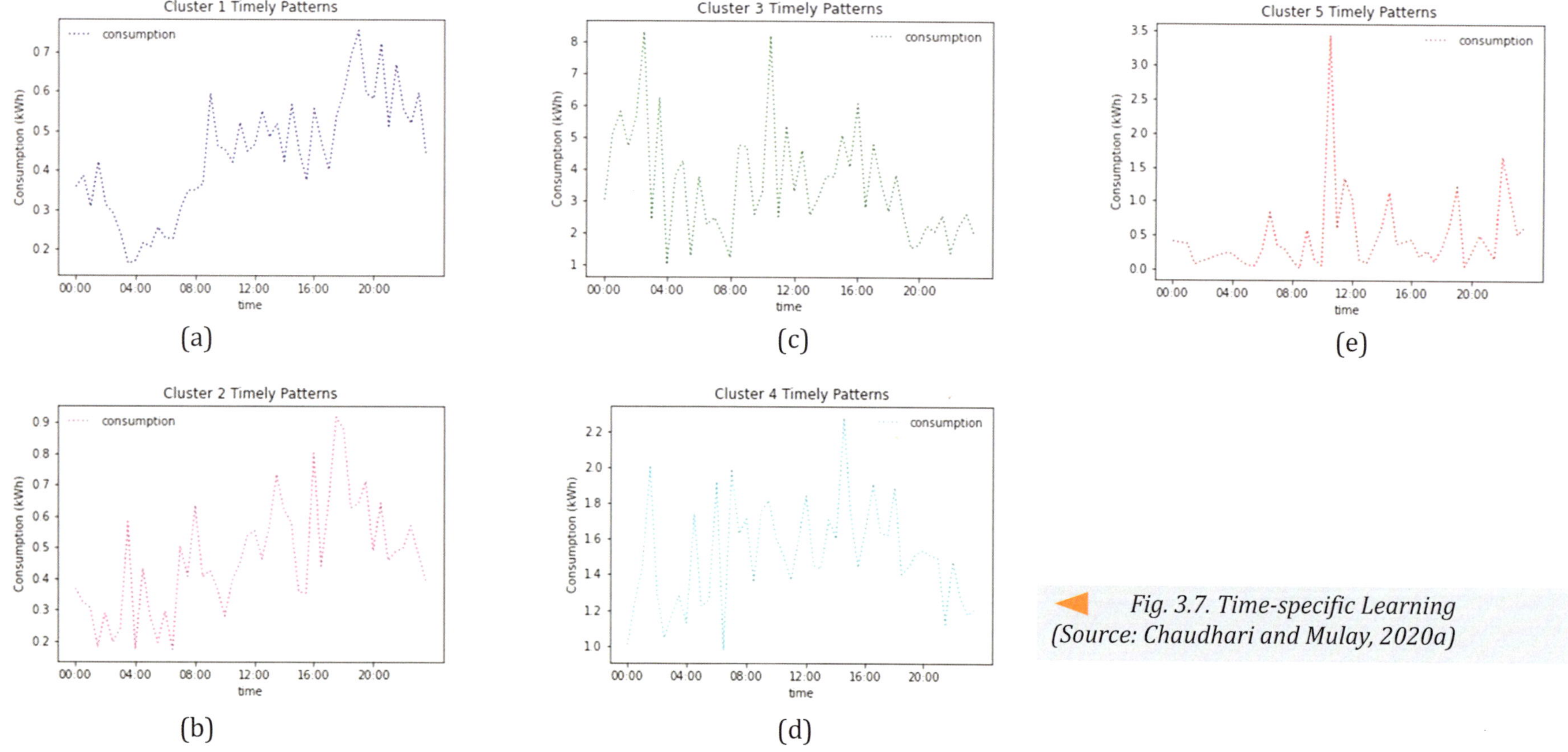

Fig. 3.7. Time-specific Learning (Source: Chaudhari and Mulay, 2020a)

The learning per cluster is:

- Cluster 1 (figure 3.7(a)) constitutes 60.73% of the population, and users in cluster 1 are average demand consumers. Interpretation of cluster 1: electricity consumption is lower during the early morning, followed by stable consumption during the afternoon and peak at about 7:00 pm.
- Cluster 2 (figure 3.7(b)) accounted for 21.58% of the population. Users in cluster 2 are ordinary consumers that consume average electricity. The peaks of electricity may appear at about 6:00 pm, after which consumption decreases significantly. Cluster 2 shows residents venturing out from home in the morning returning toward the evening and consuming electricity until they rest (go to sleep).
- Cluster 3 (figure 3.7(c)) constituted 5.98% of the population, and users in cluster 3 are highest demand consumers. They consume as much as electricity as they need without considering the price.
- Cluster 4 (figure 3.7(d)) accounted for 10.42% of the population. It contains the high demand consumer. The behavior of cluster 4 is quite similar to the cluster 3 even though it has lesser electricity consumption.
- Cluster 5 (figure 3.7(e)) constituted 1.32% of the population. Users in this cluster are low-demand consumers who consume the least electricity. Daily consumption is very stable. The peaks of electricity appear at about 7:00 am.

Season-wise Learning

- Cluster 1 shows high consumption during winter months due to more use of heaters.

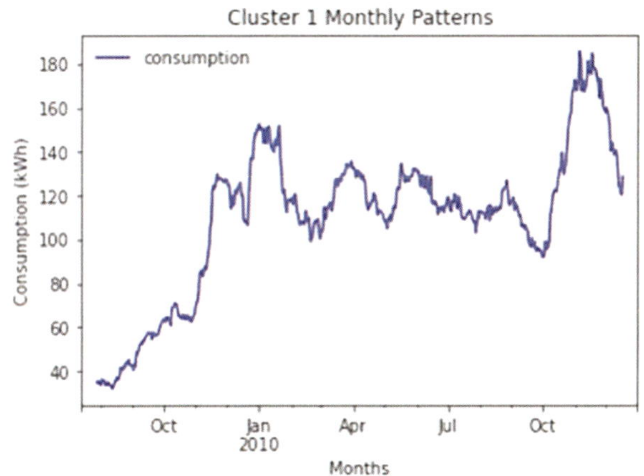

- Cluster 2 shows most consumption over the autumn months because of a very warm week in September.

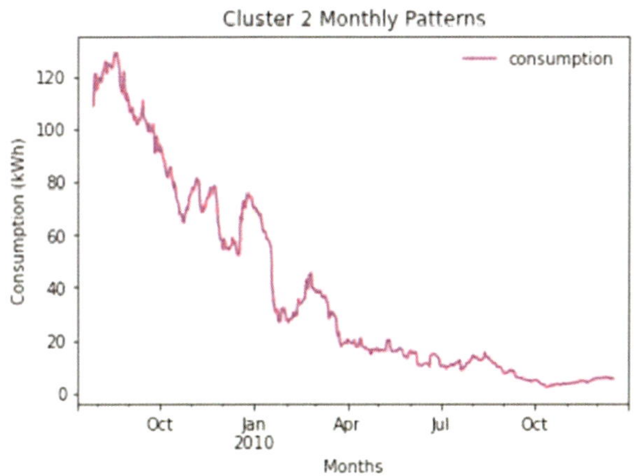

- Cluster 3 consumes most of the electricity during the winter and spring months.

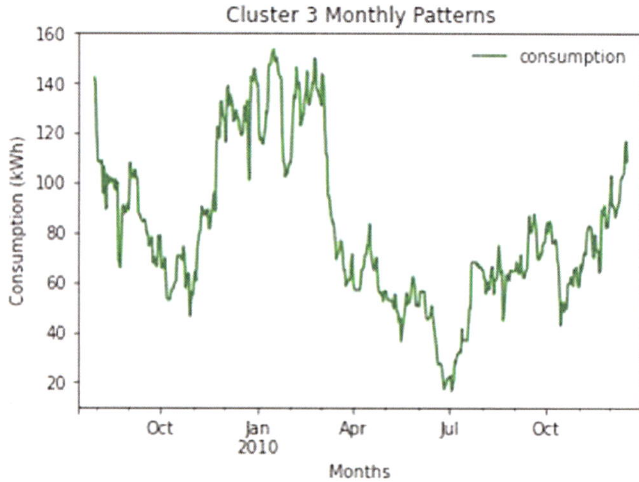

- Cluster 4 behaves in precisely the opposite manner, i.e., more electricity is consumed during spring months.

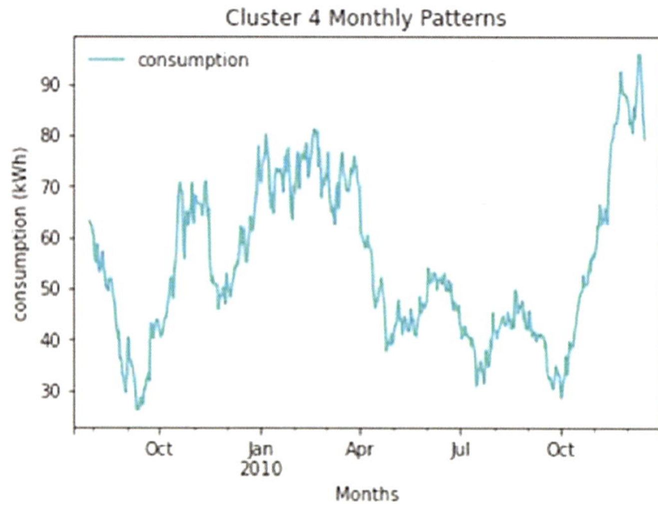

- Cluster 5 have a more uniformly constant energy profile, with the same pattern as Cluster 3.

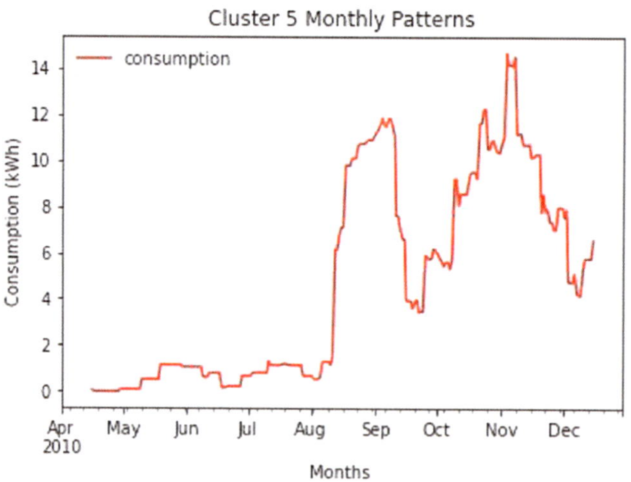

Socioeconomic Learning

- The clustering results given by CGMIC algorithm indicate a close relationship between residents' daily living conditions and the pattern of electricity consumption.
- A measurable valuation of the above observation can be made by validating the clustering results with socioeconomic data (ISSDA, 2018).

This figure 3.8(a) depicts the clusters according to social classes: low-class, middle class, and higher class. Cluster 4 contains a more significant number of high-class customers (spending limit, electricity consumption, number of appliances, family size, etc. is more) as compared to others.

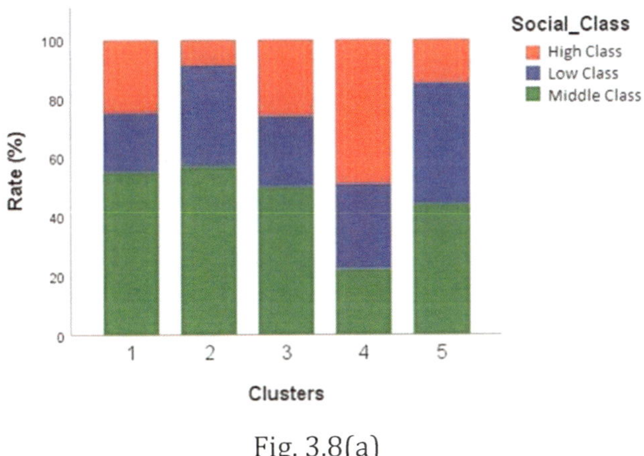

Fig. 3.8(a)

This figure 3.8(b) shows the stacked bar graph using green, red and blue colors, grouping customers according to employment status (employed, unemployed and retired) in each cluster. The number of retired persons is slightly more in cluster 3.

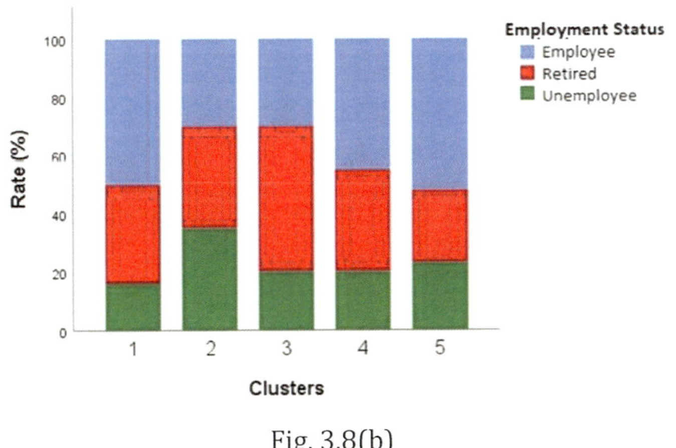

Fig. 3.8(b)

This figure 3.8(c) shows the stacked bar graph using brown, pink and yellow colors, constituting number of appliances at household residents of each cluster. Fewer appliances are used by residents in cluster 5.

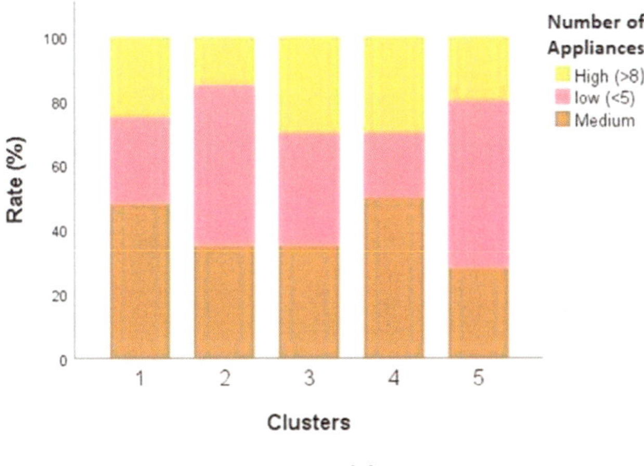

Fig. 3.8(c)

This figure 3.8 (d) shows the stacked bar graph using green and blue, indicating the usage of new technologies such as internet/Wi-Fi routers, increased usage of handheld devices, etc. It also depends on the age of residents (less age = more electricity consumption).

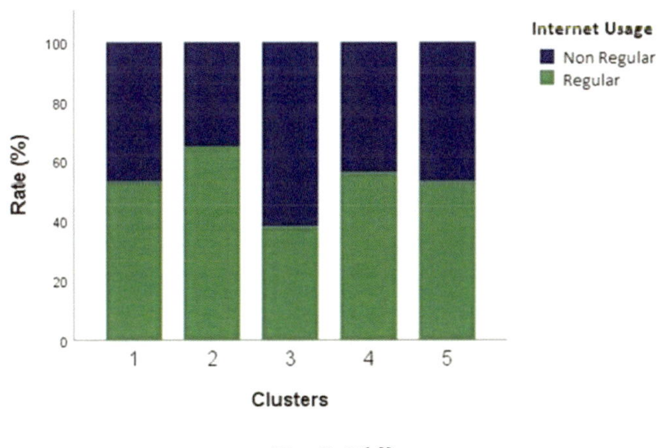

Fig. 3.8(d)

Cluster2 shows internet usage is more because it contains more young group residents.

This figure 3.8 (e) shows the stacked bar graph using red and blue. It is remarkable that the majority of consumers in all clusters use non-electric means of heating system to heat water in their home.

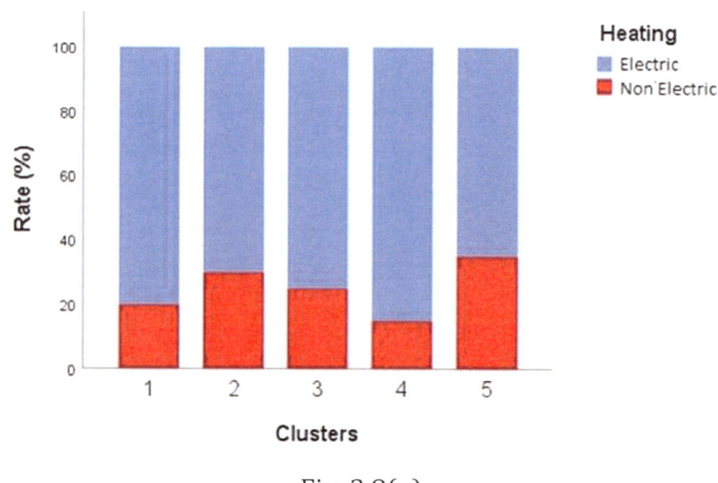

Fig. 3.8(e)

Fig. 3.8. Clustering results with socio-economic characteristics

3.6 Case Study 2: Experimental Results (Open Power System Germany Dataset)

This section explore analysis of electricity consumption and production in Germany have varied over time using proposed system.

3.6.1 Dataset Description

- Open Power System Data is a free-of-charge data platform dedicated to electricity system researchers (Wiese et al., 2019).
- Table 3.1 outlines the Germany-wide totals of electricity consumption, wind power production and solar power production for 2006-2017 (OPSD: Open Power System Data, 2020).
- Electricity production and consumption are reported as daily totals in gigawatt hours (GWh).

Germany's Electricity Dataset	
Variables	**Description**
Country	Germany
Supplier	Open Power System Data
Final no. of customers and Instances	Country-wide, 43,840
Monitoring period	1 Jan 2006 to 31 Dec 2017
Resolution	Daily
Building Type	Commercial
Attributes	Time stamp, consumption, Wind, Solar, Wind + Solar

Table 3.1: Germany's Electricity Dataset

3.6.2 CGMIC Algorithm Implementation Result on OPSD

- Figure 3.9 depicts the clusters obtained after the execution of CGMIC algorithm on the OPSD Germany dataset. The CGMIC uses closeness value and Gaussian mixture to automatically decide cluster members and number of clusters using probability and error-based computing.
- Figure 3.9 (a) presents the implementation of the CGMIC algorithm, which results in five basic clusters formed.
- We observe from figure 3.9 (b) and 3.9 (c) that on the arrival of the influx of new data, existing clusters are recomputed and new clusters are formed.

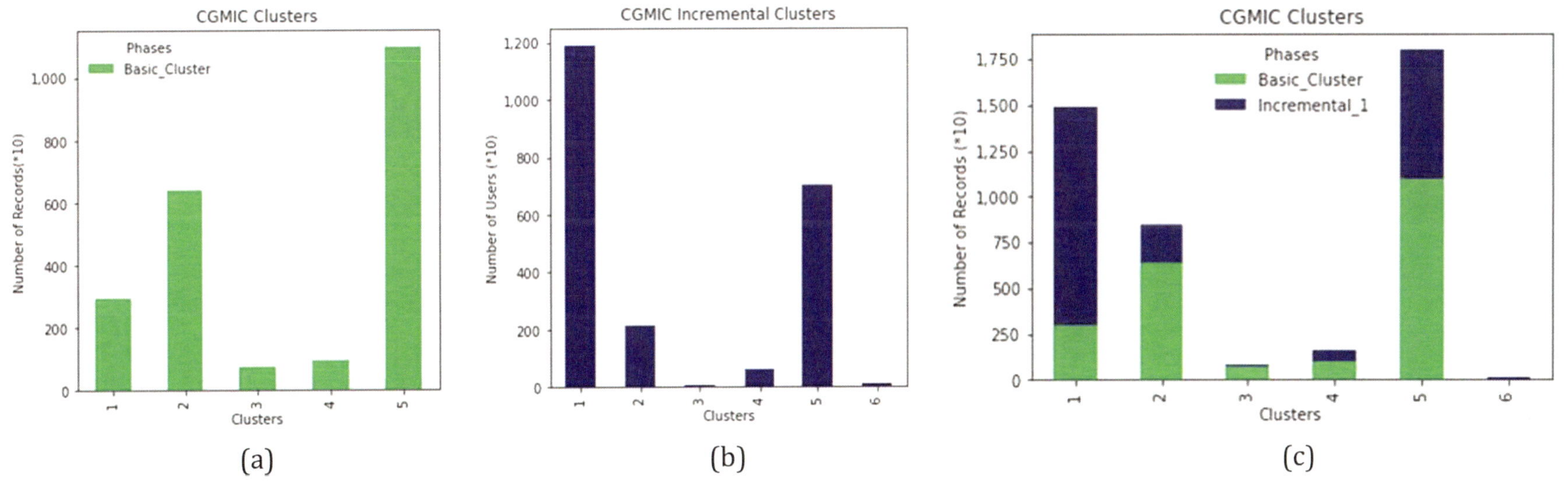

Fig. 3.9. CGMIC Algorithm Implementation Result on OPSD: (a) Describes the basic clusters; (b) Formed new cluster on influx of new data; (c) CGMIC clusters

The pattern of consumption is summarized in figure 3.10 for better illustration of the consumption behavior of different clusters. Cluster 4 consists of consumers whose electricity consumption is high, while cluster 5 consists of low-demand consumers who consume very less electricity.

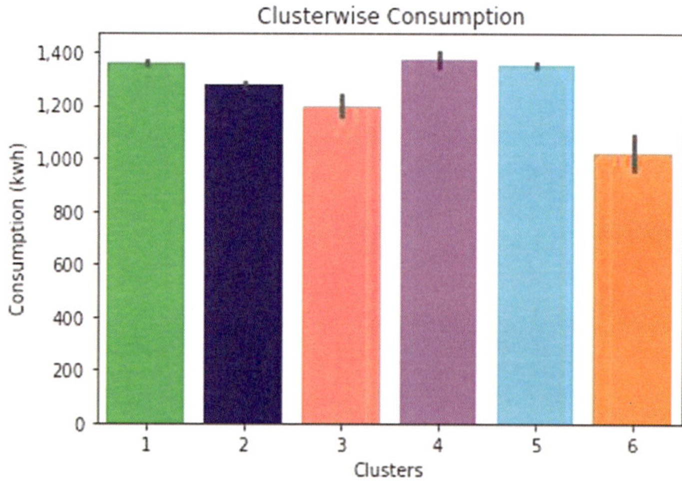

Fig. 3.10. Trends in Pattern of Electricity Consumption on OPSD 

3.6.3 Clustering Analysis

- The cluster analysis mainly aims at discovering structures in large datasets. This clustering analysis shows the trends in electricity consumption and production, typically the highest and the lowest.
- Clustering Analysis has done the following :
  - Seasonality Analysis: **Yearly**
  - Seasonality Analysis: **Monthly**
  - Seasonality Analysis: **Weekly**

Seasonality Analysis: Yearly

- Figure 3.11 (Walker, 2020) shows long-term trends in electricity consumption, solar power and wind power.
- Learning from figure 3.11 :
  - All the consumers have the highest electricity consumption and wind power production in winter and lowest in summer.
  - Solar power production shows the opposite trend, i.e., highest in summer due to abundant sunlight and lowest in winter.

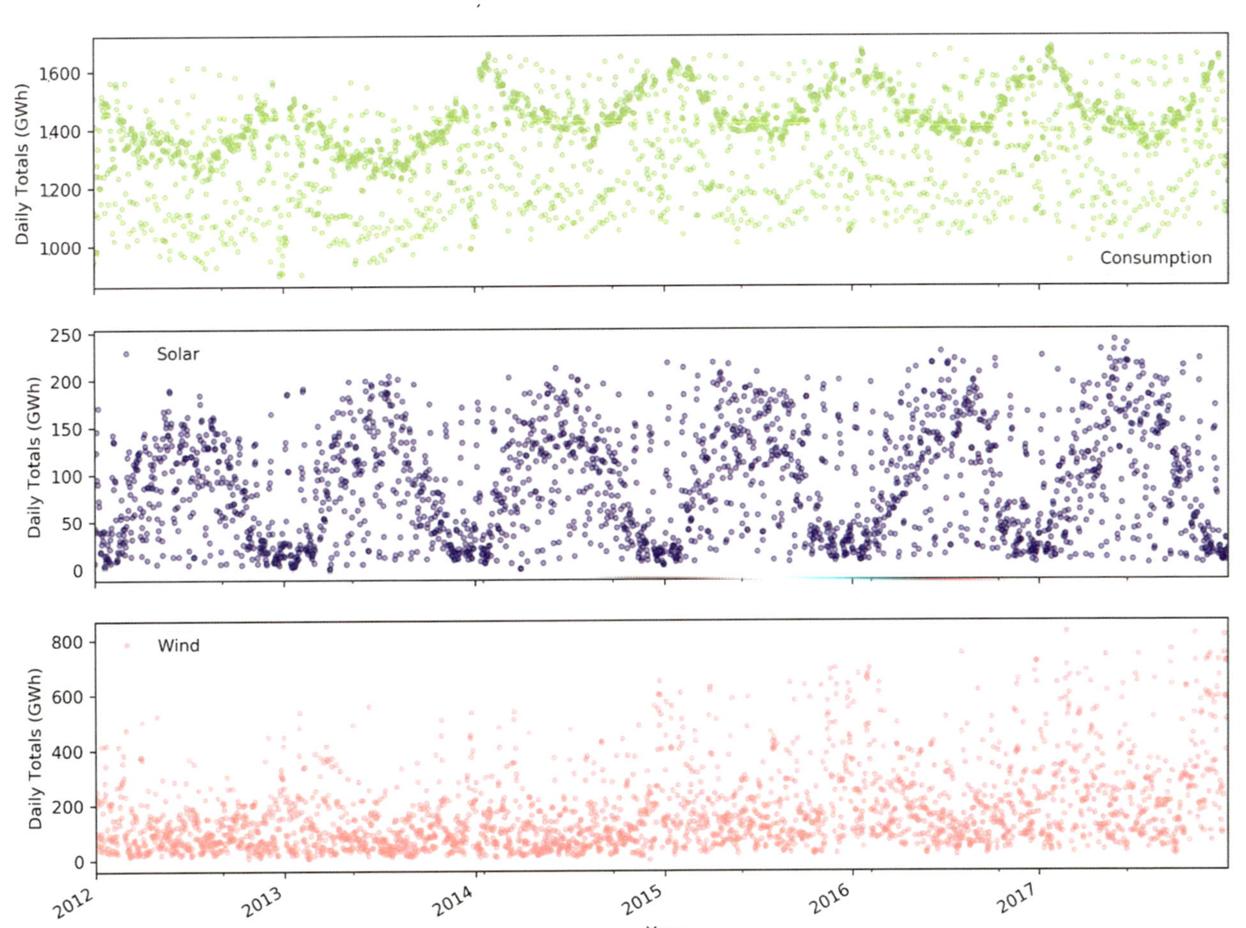

Fig. 3.11. Electricity Consumption, Solar Power, and Wind Power Yearly Seasonality Analysis

Figure 3.12 (Walker, 2020) depicts that electricity consumption has been fairly stable over time, while wind power production has been growing steadily, with wind + solar power comprising an increasing share of the electricity consumed.

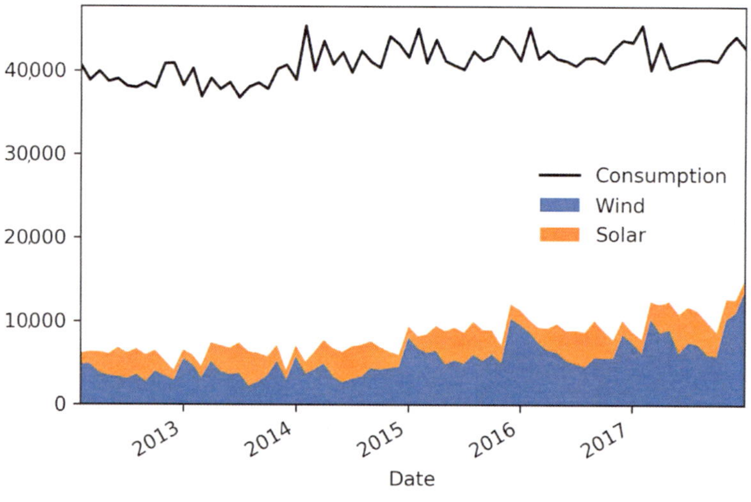

Fig. 3.12. Yearly Seasonality Analysis

Seasonality Analysis: Monthly

The box plots (figure 3.13) provide the monthly seasonality, with additional insights as follows:

- Although electricity consumption is generally higher in winter and lower in summer, the median and lower two quartiles are lower in December and January compared to November and February, likely due to businesses being closed over the holidays. This box plot confirms a consistent pattern over the years.
- While solar and wind power production both exhibit a yearly seasonality, wind power distributions have many more outliers, reflecting the effects of occasional extreme wind speeds associated with storms and other transient weather conditions.

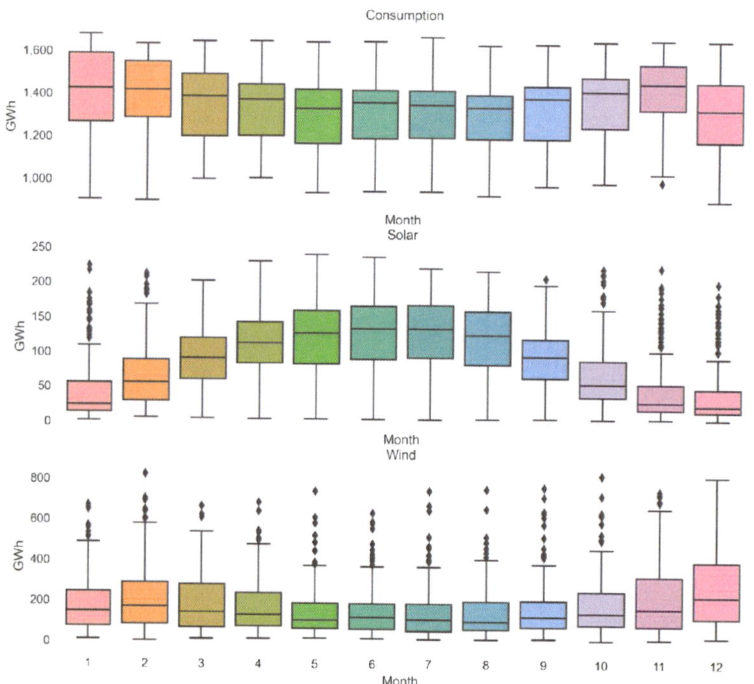

Fig. 3.13. Month-wise Learning of Electricity Consumption, Solar and Wind Power

Seasonality Analysis: Weekly

Figure 3.14 (Walker, 2020) shows weekly seasonality in Germany's electricity consumption, corresponding with weekdays and weekends. The plot suggests that electricity consumption is significantly higher on weekdays than on weekends for all the clusters. The low outliers on weekdays are presumably during holidays.

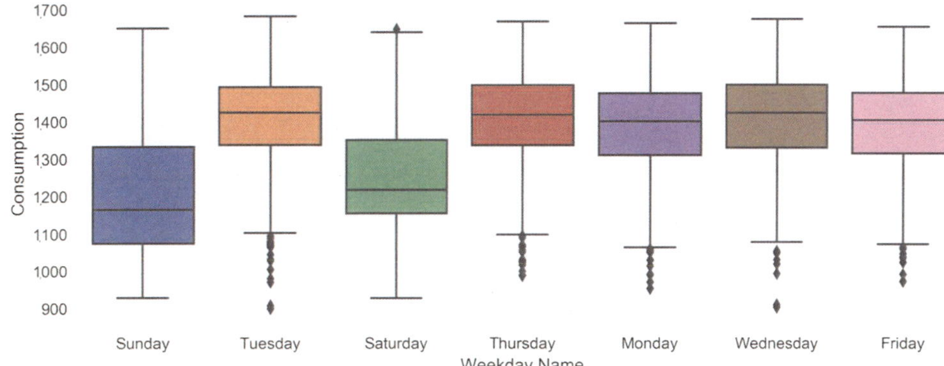

Fig. 3.14. Day Specific Learning of Electricity Consumption

Figure 3.15 (Walker, 2020) depicts the weekly oscillations in September-October 2017 electricity consumption. The plot shows consumption is highest on weekdays and lowest on weekends.

Fig. 3.15. Sept-Oct 2017 Electricity Consumption Pattern

3.7 Case Study 3: Experimental Results (IIT Bombay Indian Residential Energy Dataset)

The results of the CGMIC algorithms on IIT Bombay smart meter datasets are shown in this section.

3.7.1 Dataset Description

- This IIT Bombay dataset contains electricity consumption data from a high-rise residential building inside the IIT Bombay campus from December 2016 to January 2018, with sampling period of seconds.
- The IIT Bombay Indian Residential Energy Dataset (Mammen et al., 2018) contains a total of 118,531 samples .
- Table 3.2 describe the Indian residential power consumption dataset. The final dataset contains Unix TimeStamp (Indian Standard Time (GMT+5.30); apartment ID; voltage of phase 1, 2 and 3; active power of all three different phases; reactive power of all three different phases; current for all three different phases; power factor of all three different phases and phase angle.

IITB Indian Residential Energy Dataset	
Variables	**Description**
Country	India
Supplier	IIT Bombay : Smart Energy Informatics Lab
No. of customers and Instances	60; 118,531
Monitoring period	Dec 2016 to January 2018
Resolution	Seconds
Building Type	Residential
Attributes	Timestamp, voltage, current, active/ reactive power, PF

Table 3.2: IITB Indian Residential Energy Dataset

Figure 3.16 depicts the overall electricity consumption pattern of the residential customers over one year. The average consumption graph shows that day begins at 12:00 am and the highest peak is at 7:00 pm. Majority of the electricity consumption occurs in the evening between 6:30 pm and 8:00 pm.

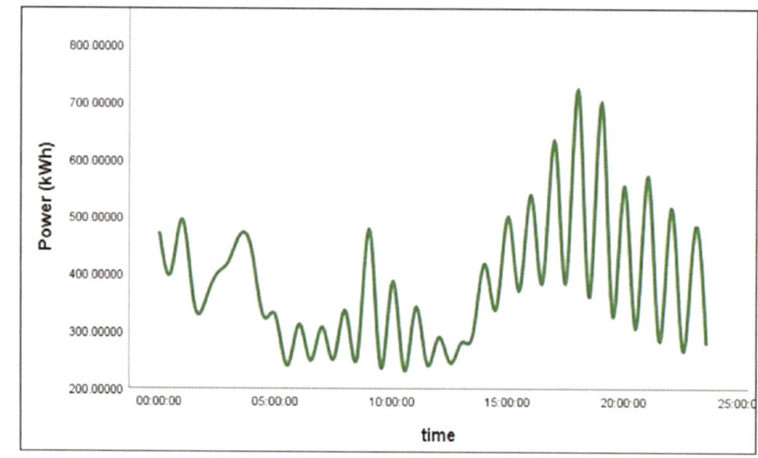

Fig. 3.16. Average Electricity Consumption of Residential Sectors for the year 2017

3.7.2 CGMIC Clustering Result on IIT Bombay

- Figure 3.17 depicts the clusters obtained from the CGMIC algorithm applied on IIT Bombay dataset.
- Figure 3.17 (a) presents phase one implementation of the CGMIC algorithm, which results in four basic clusters formed. The influx of new data fits into the existing cluster as seen in figure 3.17(b).
- We observe from figure 3.17(c) and 3.17(d) that on arrival of an influx of new data, some of the data do not fit into existing clusters because closeness value is different, so the CGMIC algorithm automatically forms new cluster numbers five and six, respectively.
- CGMIC divides 1 lakh ESM data into five clusters. Consumers are grouped based on electricity usage patterns/behaviours.

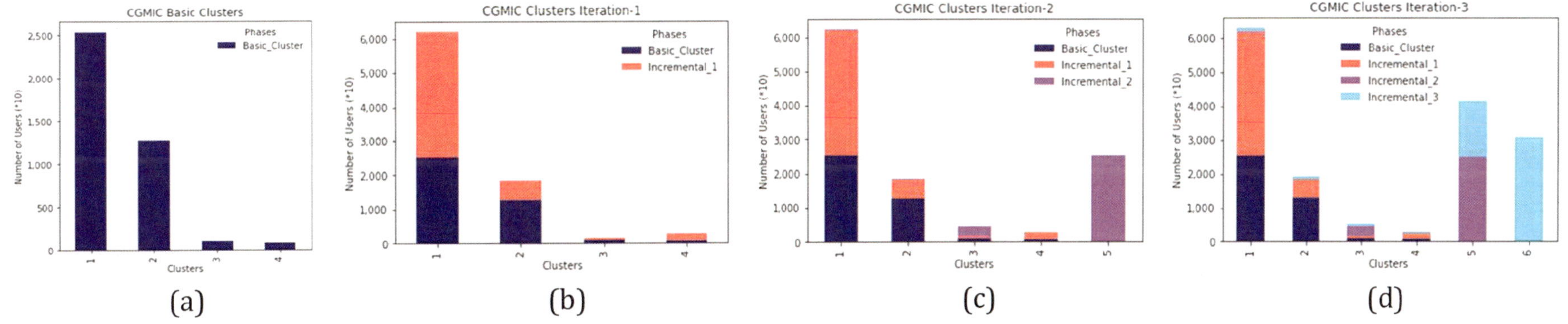

Fig. 3.17. CGMIC Clustering Results on IIT Bombay dataset: (a) Describes the basic clusters; (b) Shows the update to existing clusters; (c) and (d) show formation of new clusters

3.7.3 Result Analysis

The cluster-wise consumption patterns of the residential customer are depicted in figure 3.18. Cluster 4 contains consumers who have consumed the highest electricity. Cluster 1 and Cluster 5 contains customers whose electricity consumption is average; that is why cluster 1 and 5 are updated frequently.

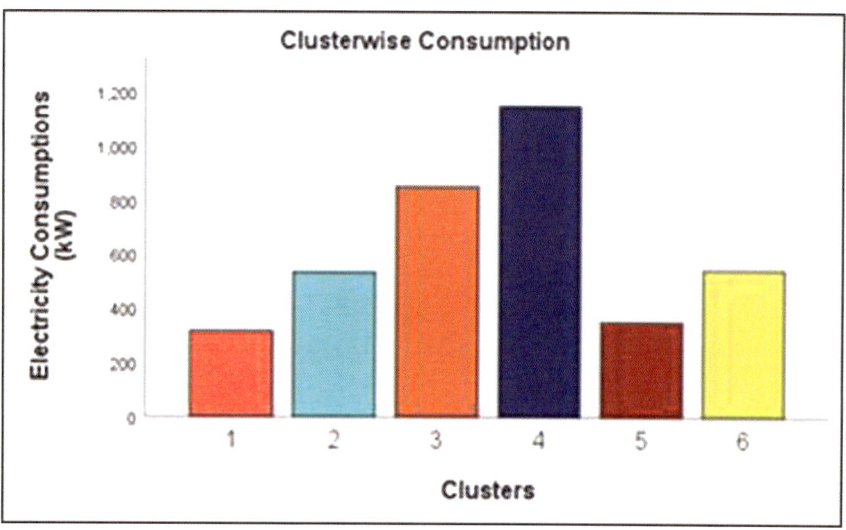

Fig. 3.18. Cluster-wise consumption analysis

Clusters	1	2	3	4	5	6
Percentage of Data	57.9 %	6.3 %	10.4 %	4.2 %	16.4 %	4.8 %

Table 3.3 : Descriptive analysis of Clusters generated from CGMIC

3.8 Case Study 4: Experimental Results (Low CarbonLondon Dataset)

This section describes household clustering focus on consumption-pattern-oriented approach. The accessibility of high resolution consumption information allows clustering of households in progressively precise and innovative manners.

3.8.1 Dataset Description

- In the Low Carbon London (LCL) project (Schofield et al., 2013), Landis and Gyr (L+G) E470 ESMs were installed in 2,639 residential households across the Mayor of London's Low Carbon Zones and the London Power Networks distribution network license, i.e., the area operated by U.K. Power Networks (Sun et al., 2018).
- The collected LCL dataset consists of 46,235,280 half-hourly measurements of demand across 2,639 customers in kW for a full calendar year from January 1, 2013 to December 31, 2013.
- Table 3.4 describes the LCL dataset. In addition, various socioeconomic data of the participating households were also recorded in the dataset.
- Table 3.5 illustrates households in the UK have been allocated to CACI CORN (Classification of Residential Neighborhoods) categories, which is a geo-demographic segmentation of UK's population (Daftari, 2018).
- Three such groups - Affluent, Comfortable and Adversity - can be determined as a rough indicator of wealth.

Low Carbon London Dataset

Low Carbon London Smart Meter Dataset	
Variables	**Description**
Country	London
Supplier	UK Power Networks
No. of Customers	2639
Monitoring period	6/12/2011 to 28/2/2014
Resolution	Half hourly
No. of instances	1 million
Building Type	Residential
Attributes	Meter id, date, time, tariff, Acorn group, weather, demographic data

Table 3.4 : LCL Dataset Description

	Household Size			
	(No. of occupants)	**1**	**2**	**3 +**
Demographic Grouping	Adverse (ACORN Groups ABCDE)	315	278	234
	Comfortable (ACORN Groups FGHIJ)	240	304	214
	Affluent (ACORN Groups KLMNOPQ)	431	400	223

Table 3.5 : Customer Categories: Demographic and Household Size

3.8.2 CGMIC Clustering Result on LCL

- CGMIC divides the consumption range into six categories based on the electricity consumption range as shown in figure 3.19.

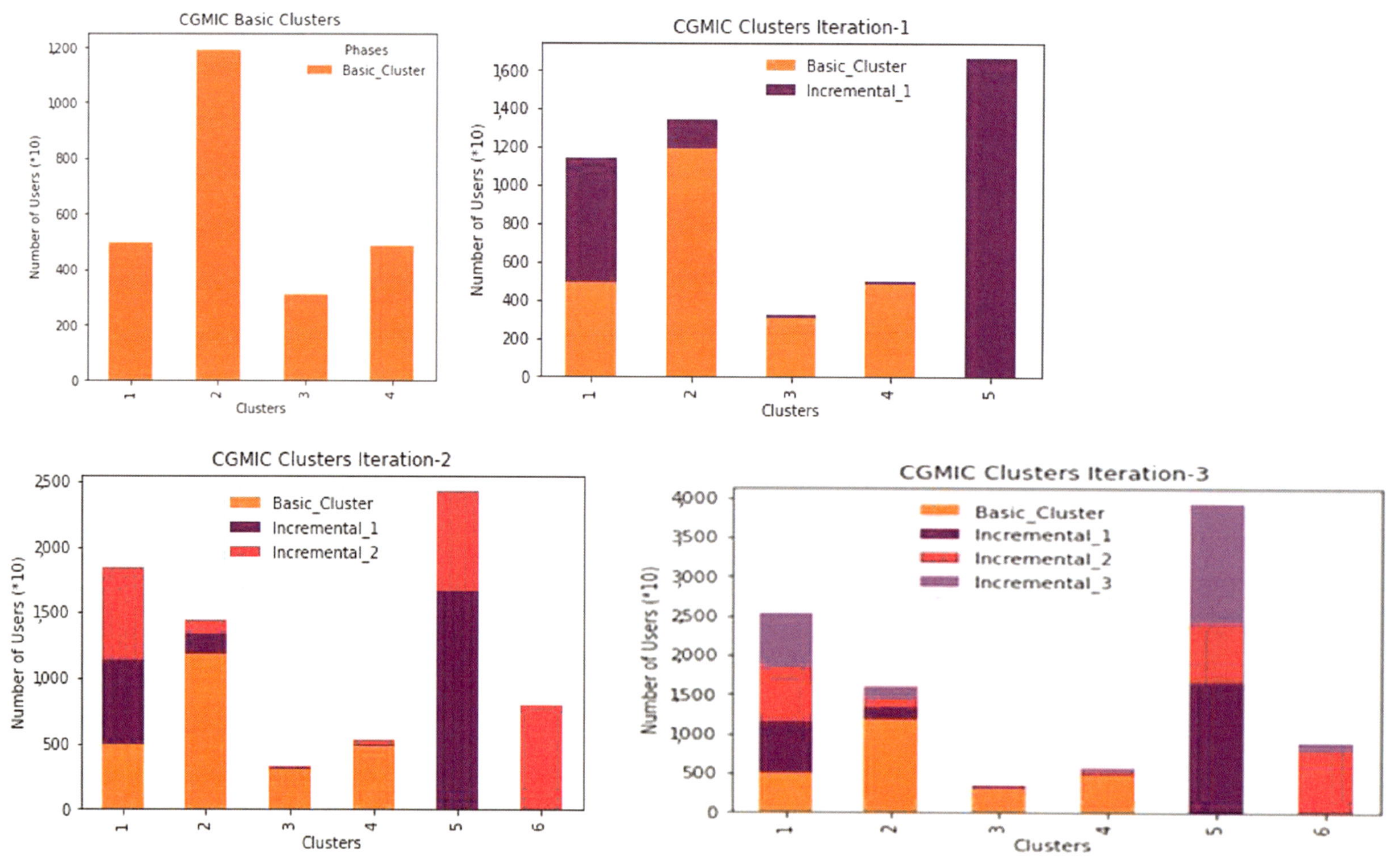

Fig. 3.19. CGMIC Clustering Results on Low Carbon London dataset : (a) Describes the basic clusters; (b) shows new clusters formed (c) shows the update to existing clusters and form new cluster (d) update existing cluster

3.8.3 Data Insights with Cluster Analysis: LCL

- There are six clusters formed after the execution of CGMIC. Consumers have been grouped based on electricity usage as shown in figure 3.20.
- Cluster 1 (figure 3.20(a)) comprises high-demand consumers and constitutes 50.61% of the population. Interpretation of cluster 1: electricity consumption is lower during the early morning, followed by stable consumption during the afternoon and peak at about 7:00pm.
- Cluster 2 (figure 3.20(b)) comprises ordinary consumers that consume average electricity, accounting for 12.30% of the population. The peaks of electricity use may appear at about 10:30am. Cluster 2 shows that a majority of the residents leave their home during the day, and consume more electricity at their lunch and evening times.

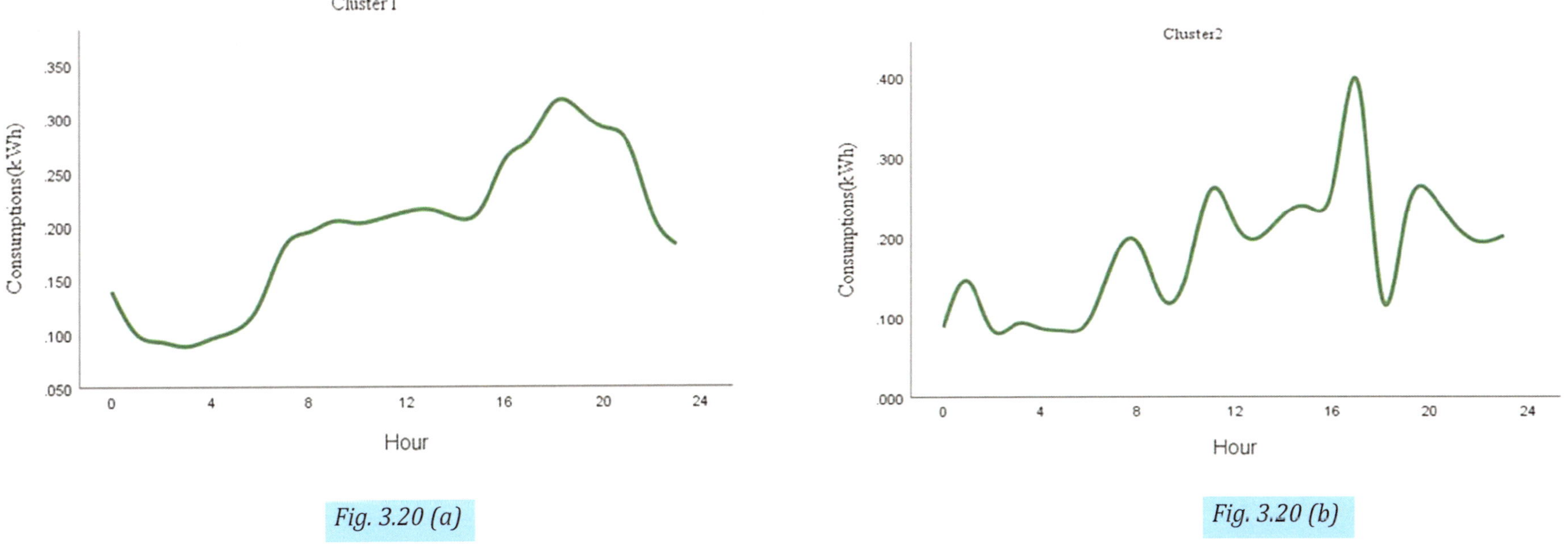

Figure 3.20. Average Electricity Consumption across the same Hours on Different Day of CGMIC Clusters

- Cluster 3 (figure 3.20 (c)) constituted 19.84% of the population. The behavior of cluster 3 is quite similar to cluster 2 although it has higher electricity consumption.

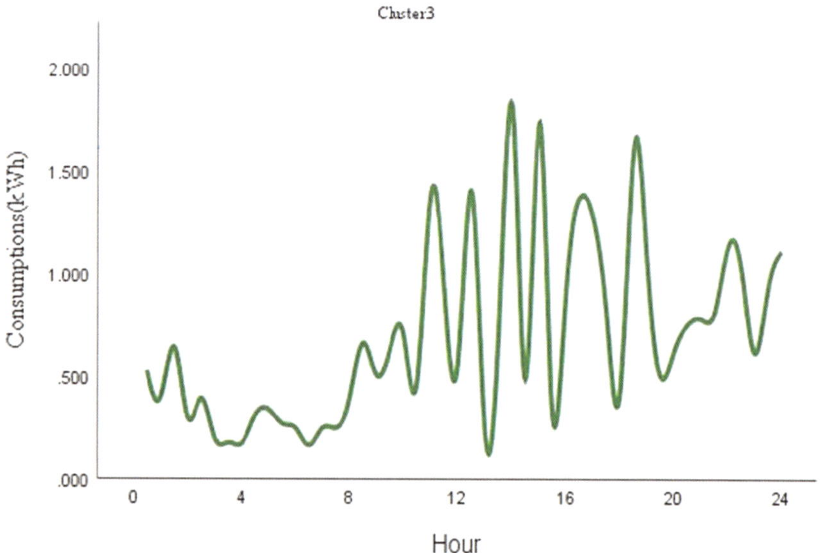

Fig. 3.20 (c)

- Cluster 4 (figure 3.20 (d)) accounted for 9.70% of the population and comprises the highest-demand consumer. The electricity consumption of users in this cluster is not stable. Users consume as much as electricity as they need without considering the price. The peaks of electricity use appear at about 7:00am.

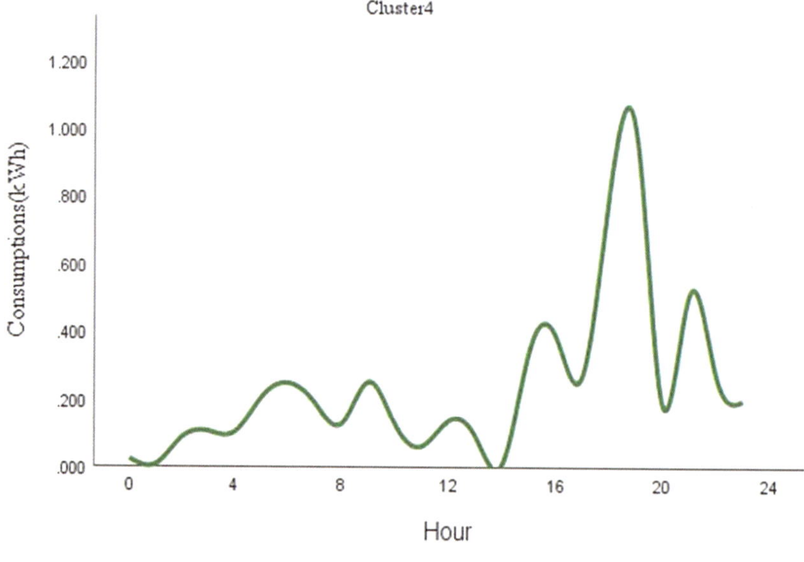

Fig. 3.20 (d)

- Cluster 5 (figure 3.20 (e)) constitutes 10.33% of the population. Users in this cluster are low-demand consumers, using the least electricity. Daily consumption is very stable, and peak consumption point at 1.29 kWh per day.

- Cluster 6 (figure 3.20 (f)) accounted for 8.23% of the population. This cluster comprises users who consume very low electricity. The patterns of consumption are similar to cluster 5.

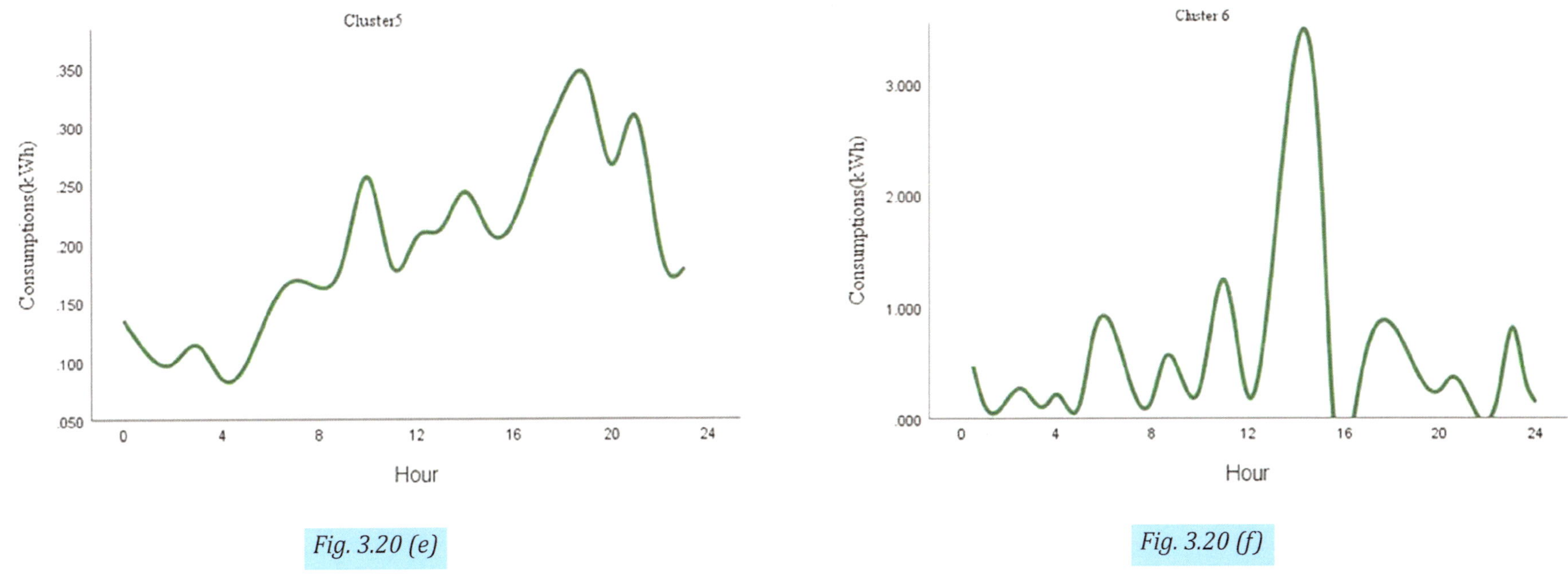

Fig. 3.20 (e)

Fig. 3.20 (f)

These cluster details are very useful for electricity utilities to reduce power outage and may help generation units reduce capital expenses of building new plants.

3.8.4 Incremental Learning of Electricity Smart Meter Data Analysis

Electricity Consumption by Different Tariff Groups

- The figure 3.21 shows the usage is very high during evening hours and less during day time. As expected, the mean energy usage dips to the lowest during late-night hours. This pattern remains the same irrespective of their tariff rates (high, normal or low). Also, it is seen that when high rates are charged, subscribers tend to use a little less energy so as to reduce their total expenses. Whereas, at lower rates, the average energy consumption is more compared to when high/normal tariff rates were applied.

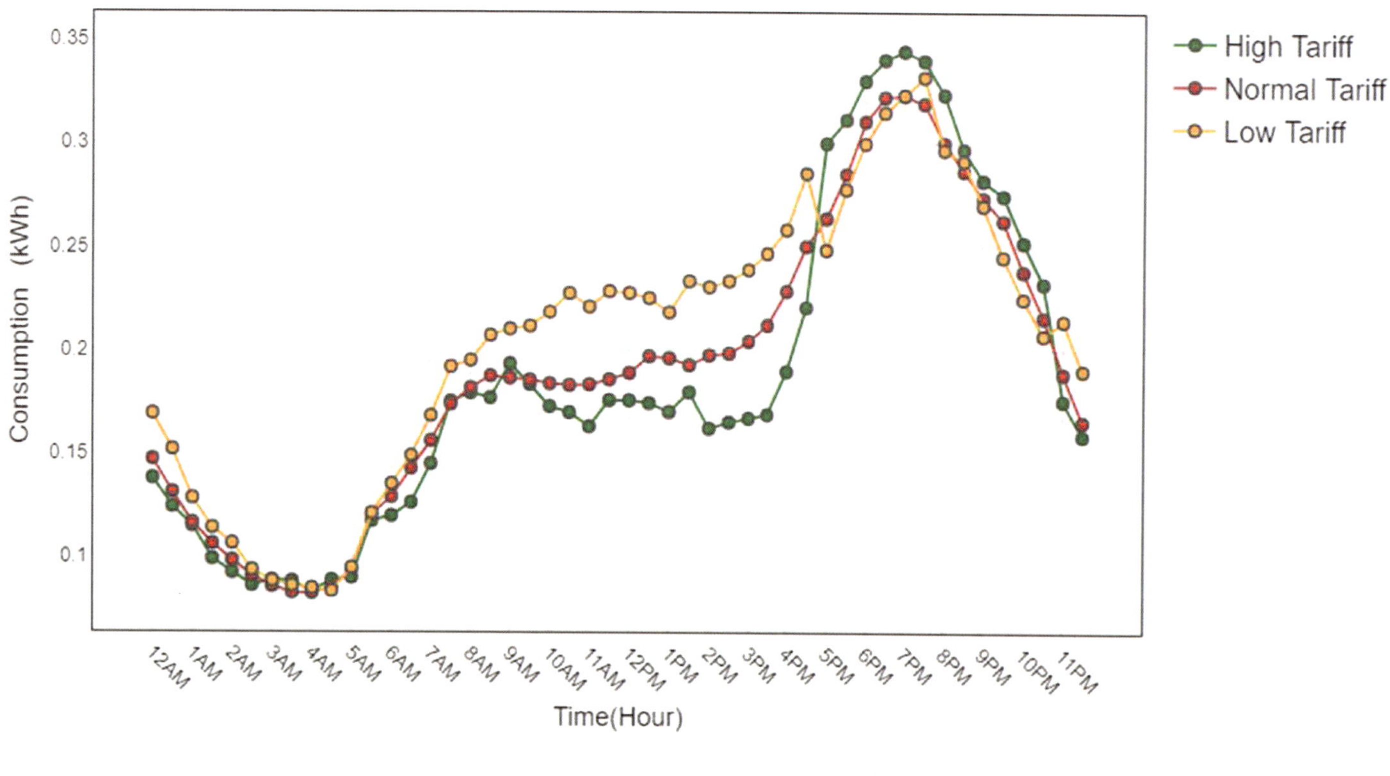

Fig. 3.21. Learning of Electricity Consumption by Different Tariff Rates

Temperature vs. Mean Electricity Consumption per ACORN Group

- Figure 3.22 depicts the average temperature for the year 2013 of month December, January, February and March was below 5°C. This explains higher electricity consumption during this period as heating systems would have been used extensively (Daftari, 2018).
- So, temperature change plays a major role in energy consumption irrespective of different ACORN groups.

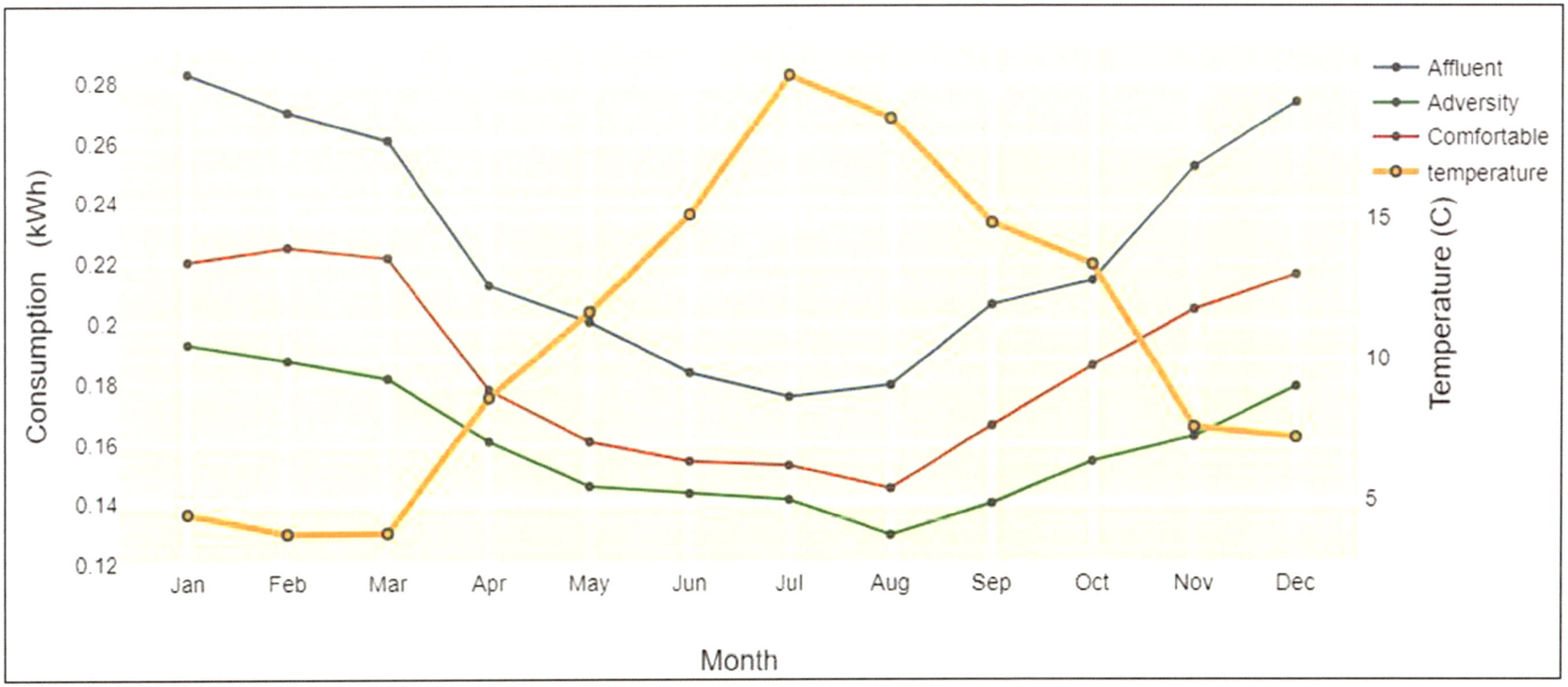

Fig. 3.22. Temperature Vs Mean Electricity Consumption per ACRON Group over the Year 2013 

Season-wise Learning

- The amount of electricity consumed by households tends to closely reflect seasonal patterns.
- Figure 3.23 shows all three Acorn groups (Affluent, Comfortable and Adversity) consumed more electricity in winter season and least electricity in summer seasons, presumably because heating systems would have been used extensively.

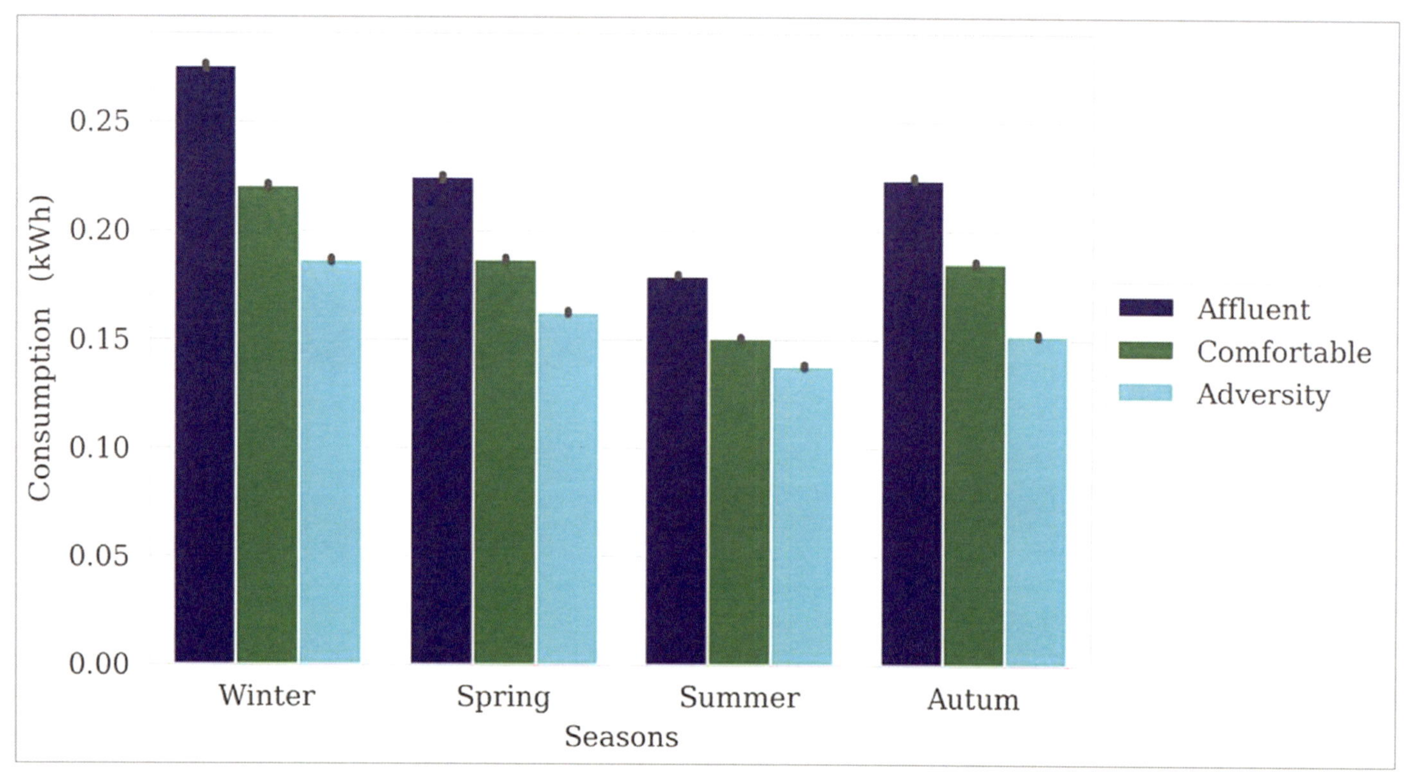

Fig. 3.23. Season-wise Learning per ACRON Group

3.9 Case Study 5: Experimental Results (Prayas Energy Group (India) Dataset)

In this section, we study the performance of CGMIC Algorithm on the Prayas Energy dataset.

3.9.1 Dataset Description

- The electricity smart meter dataset in the Pune (India) area provided by Prayas Energy Group, Pune, India as shown in table 3.6.
- The period of this particular dataset is from 1st Nov 2018 to 2nd Nov 2018 (Prayas Energy Group, 2018).
- The dataset contains minute-level electricity consumption data from 70 houses.
- The ESM dataset contains more than 1 lakh samples.

Prayas Energy Group Dataset	
Variables	**Description**
Country	India (Pune)
Supplier	Prayas Energy Group
Initial no. of customers	70
Monitoring period	1 November 2018
Resolution	1 minute
Building Type	Residential
Attributes	Meter ID, date, time, commutative usage(kWh), and voltage

Table 3.6: Prayas Energy Group Dataset

3.9.2 CGMIC Clustering Result Prayas Energy Group (India)

- Figure 3.24 depicts the clusters obtained after the execution of CGMIC on the Prayas dataset. The CGMIC uses closeness value and Gaussian mixture to automatically decide cluster members and number of clusters using probability and error-based computing.
- Figure 3.24 (a) presents the implementation of the CGMIC system, which results in four basic clusters formed. We observe from figure 3.24 (b) that on the arrival of the influx of new data, existing clusters were recomputed.
- Some of the influx of new data fit into the existing clusters and the remaining formed new clusters as shown in figure 3.24 (c).

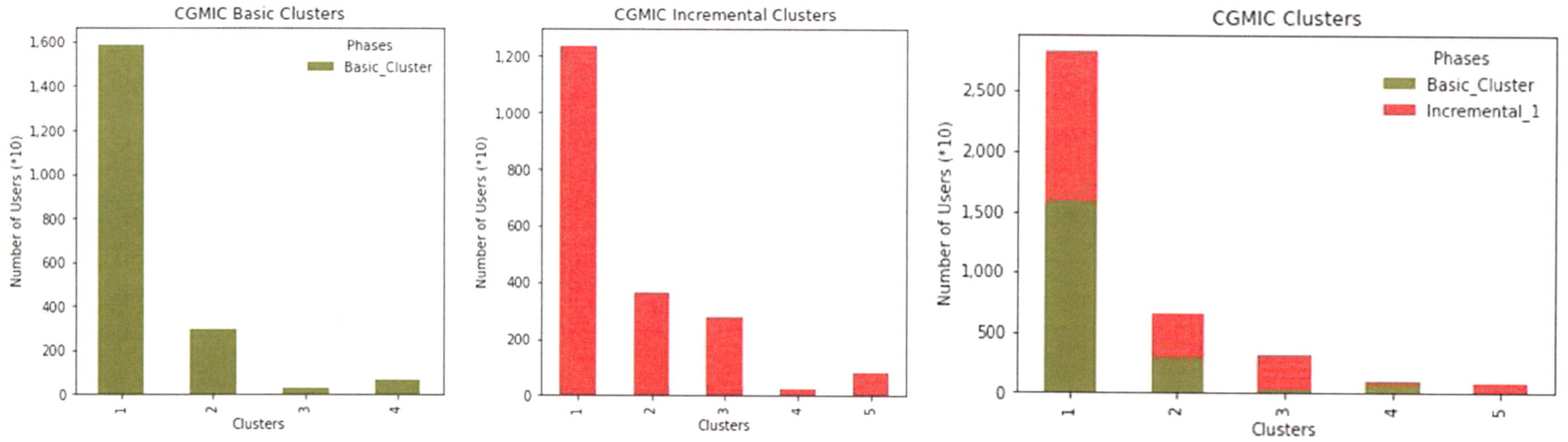

Fig. 3.24. CGMIC Clustering Results on Prayas (Indian) dataset : (a) describes the basic clusters; (b) shows incremental clusters; (c) shows update in the existing clusters and new clusters formed

3.10 Result Validations

- It is important to validate any clustering found in order to determine if the clustering is correct or not, and if the dataset even contains some form of clustering tendency.
- To validate results given by the CGMIC algorithm, professional EM and DBSCAN (Ester et al., 1996) algorithms are used in this research.

3.10.1 Cluster Evaluations

The cluster generated by the CGMIC algorithm is compared with the predefined clustering algorithm for validation (Table 3.7). It is worthy of note that CGMIC converges faster and may achieve better converge point compared to that with concurrent updates due to distributed implementation with loss-free optimization.

Characteristics	CGMIC	EM	DBSCAN
Cluster First Approach	Yes		Yes
Parameter-free	Yes		Yes
Converge Guaranteed	Yes	Yes	Yes
Data structure independent cluster analysis	Yes	Yes	
Outlier Formations	Yes		Yes
Order Independence	Yes		Yes
User input required		Yes	
Incremental learning	Yes		Yes
Cluster and attribute ranking during iterations	Yes		Yes
Time/day specific learning	Yes		Yes
Seasonal electricity consumption learning	Yes		Yes
Planned and unplanned outage computation	Yes		Yes

Table 3.7: Algorithmic and Cluster-oriented Comparative Details

Source: Kulkarni and Mulay (2013)

3.10.2 System Performance Evaluations

- In power system literature, Davies-Bouldin index (DB), Dunn validity index (DVI) and Silhouette Width Criterion (SWC) (Jiang et al., 2015; Davies and Bouldin, 1979) are among the most popular indices and are therefore used in this study too.
- Table 3.8 shows the clustering performance comparison of the CGMIC algorithms with the predefined clustering algorithm DBSCAN on the same dataset. As per definitions of DB, DVI, and SWC, CGMIC outperforms the DBSCAN and EM methods. Hence, the proposed CGMIC method is more robust than the DBSCAN. Smaller values of DB imply that the clustering algorithm separates the dataset properly. Hence, the method presented here generates better clustering results.

Methods	CGMIC	EM	DBSCAN	Bounded Score*
DB-	0.1090	1.1985	1.6529	0 to 1
DV I +	1.1873	0.6584	0.1002	An optimal value of the k maximizes the DVI
SWC +	0.3654	0.4078	1.2794	-1 to 1

**(Arbelaitz et al., 2013)*

-: minimum is the best; +: maximum is the best

Table 3.8: Clustering Performance Evaluations

3.11 Cloud4CGMIC Validations

The Cloud4CGMIC system was successfully implemented on VMs (compute units) of Microsoft Azure. Microsoft Azure automatically selects the numbers of VMs for implementing the applications. Hence, CGMIC runs on six VMs, of which one unit of small size is allocated for web role and the remaining are used for worker role in testing.

3.11.1 Time Complexity

- Figure 3.25 proves that time required to execute (Ireland Smart Meter dataset) static (desktop version) CGMIC is more as compares to Cloud4CGMIC (distributed version).
- Microsoft Azure provides the necessary processing power for the distributed version of the algorithm.

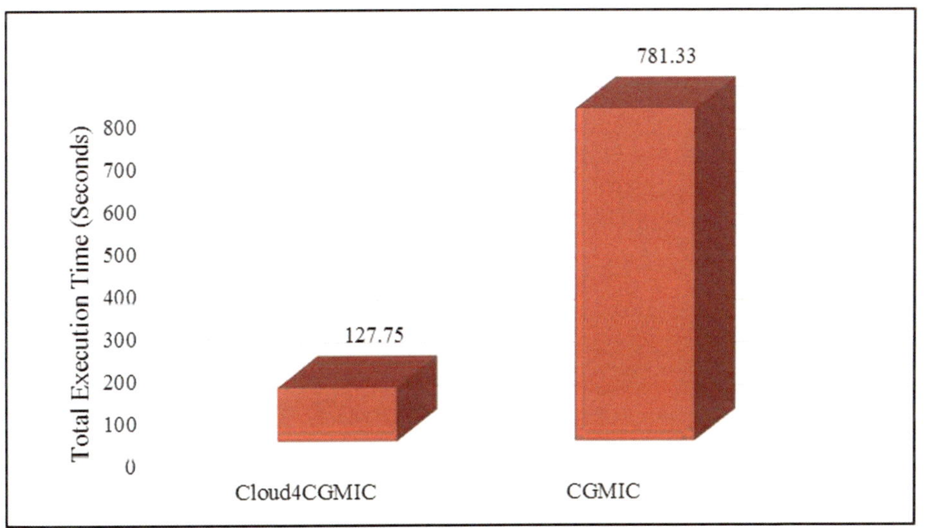

(Source: Chaudhari and Mulay, 2020b)

Fig. 3.25. Execution Time Required by Static and Distributed CGMIC

3.11.2 Scalability

- The goal of the scalability experiments is to determine the effects of the number of nodes in the system on the execution times.
- There are two types of scalability: horizontal scalability (number of worker role instances) and vertical scalability (size of worker role instances). All tests were performed on the Microsoft Azure cloud.
- Figure 3.26 shows the total execution time required by the instances of the worker role. Figure 3.27 depicts the speed of each instance of the worker role.

Fig. 3.26. Result of Horizontal Scalability on Both Case Studies

Fig. 3.27. Processing Speed of Instances of Worker Role on Both Case Studies

- It proved that total execution time for pattern recognition exponentially reduced as number of computing unitd increased by 8 to 16 instances. Thus, the algorithm can comfortably handle high-dimensional data because of its low complexity.
- Figure 3.28 shows that execution of a single task on the dataset was performed faster on S-size instance and slower on XL-size instance. However, S-size instance can execute only one task, and XL-size instance can execute eight load prediction tasks at the same time (Mrozek et al., 2015).

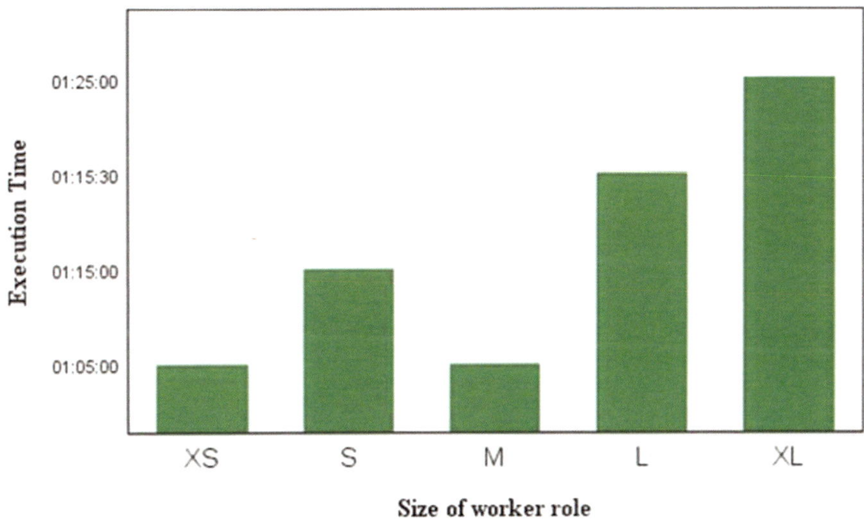

(Source: Chaudhari and Mulay, 2020b)

Fig. 3.28. Result of Vertical Scalability on both case studies

3.12 Summary of Chapter 3

- ESM generates a wealth of consumption as well as related socioeconomic data dynamically. However, the huge amount of data and the number of data types involved make data analysis highly complex, which requires dynamic systems.
- Microsoft Azure provides the processing power necessary to handle data analytics in a pay-as-you-go model. The developed Cloud4CGMIC combines ESM, cloud storage, and algorithmic data analysis to unlock efficiency in the national energy grid.
- Cloud4CGMIC platform is scalable, can do manual or auto load balancing if one of the instances of manager role and worker role goes down, and concurrency of operations is achievable.
- Queues in Cloud4CGMIC are reliable storage for closeness value calculation, and pattern recognition does asynchronous communication and maintains load balancing. Total execution time for pattern recognition is proved to exponentially reduce as a number of computing units increases.
- The Cloud4CGMIC not only helps in lower electricity consumption but also helps reduce carbon emission.
- The findings of this study will help:
- Household customers to monitor and improvise electricity consumption patterns.
- Utility providers to reduce power outage and capital expenses of building new plants.

Chapter 4

Load Forecasting using LSTM Network

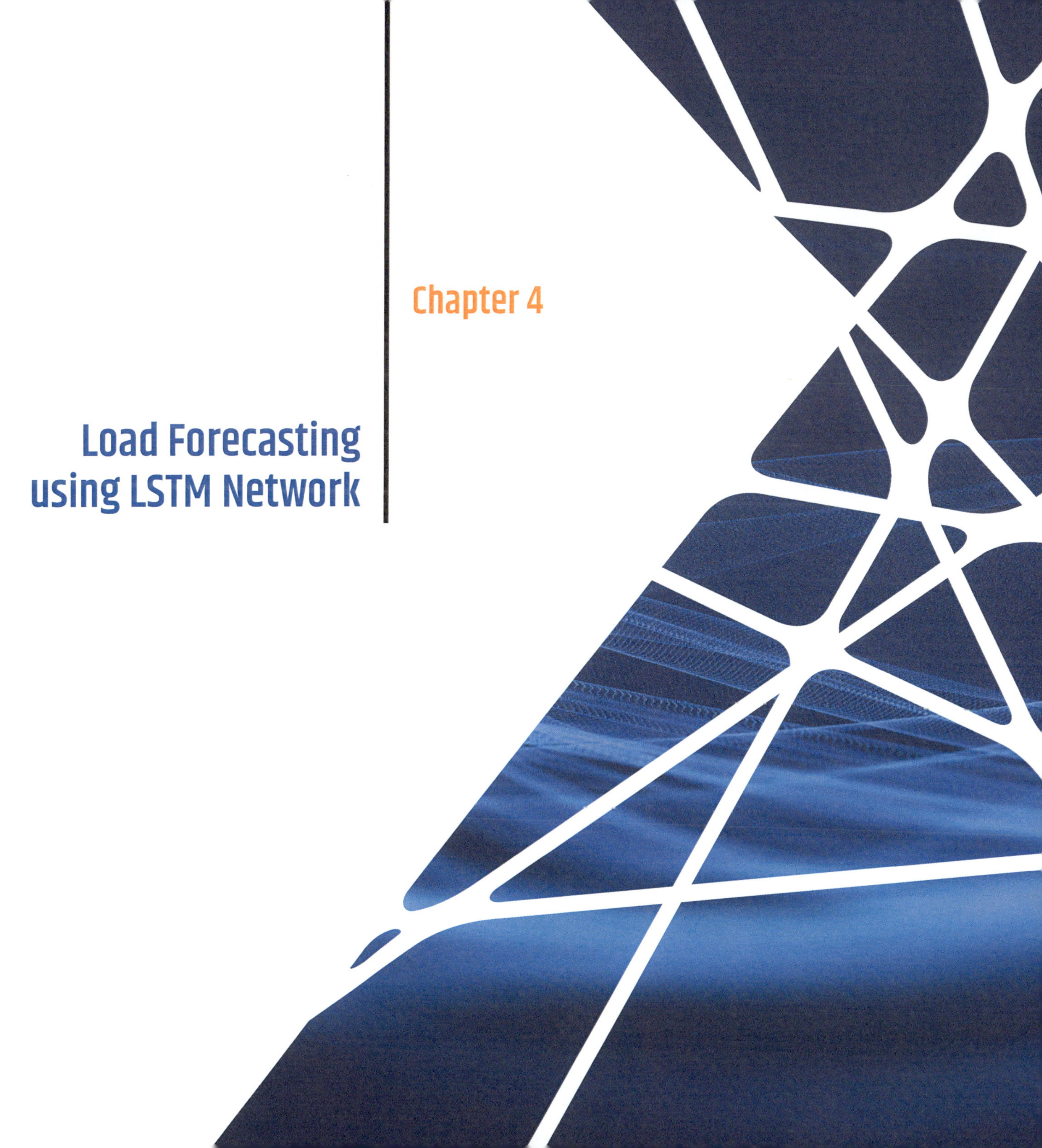

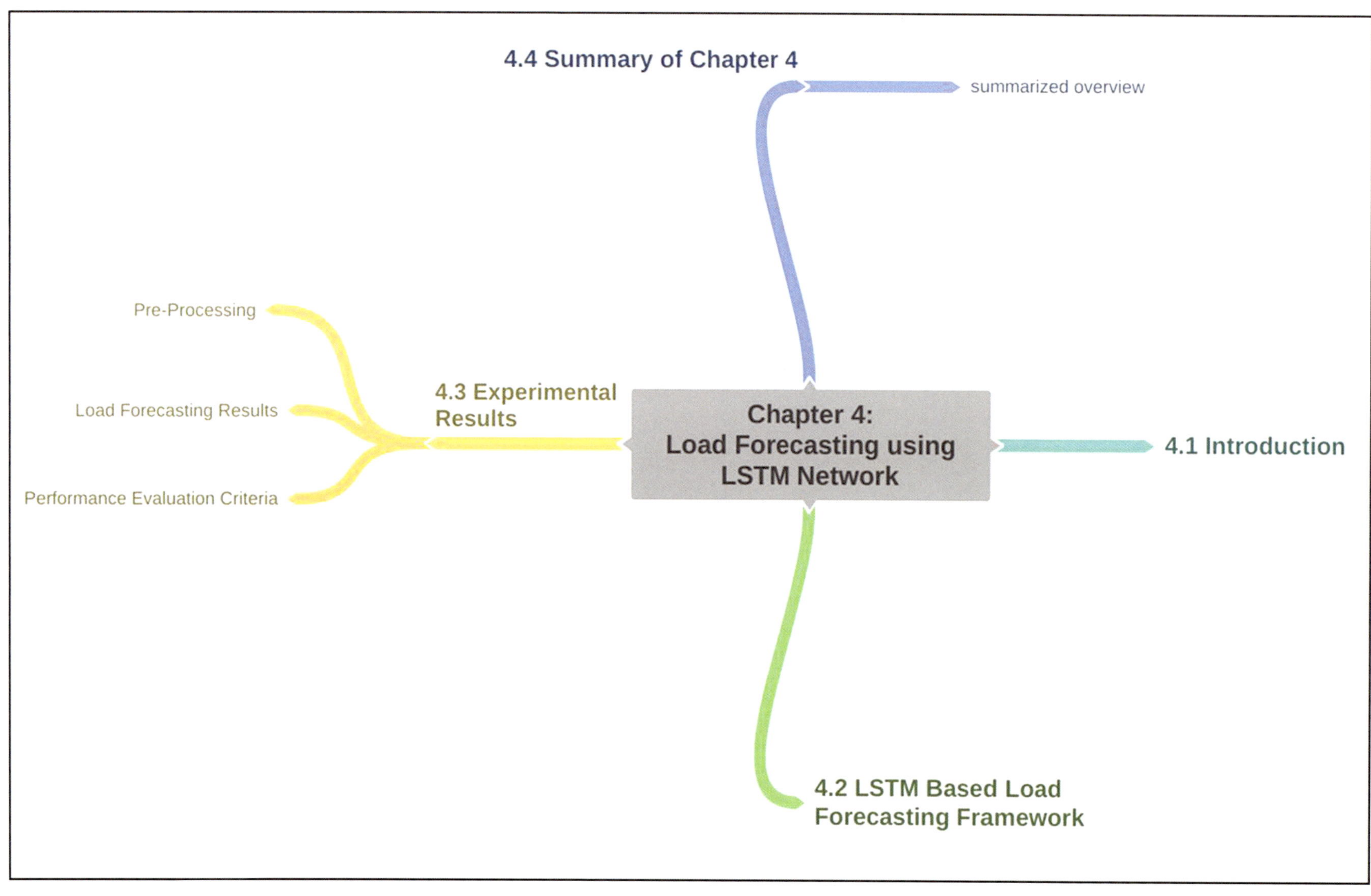

Mind map of Chapter 4

Source: https://coggle.it/

4.1 Introduction

- This chapter presents a smart electricity meter load forecast of the target time interval using the Long Short-Term Memory (LSTM) network.
- Load forecasting anticipates the amount of power needed to supply the demand and is an important component of power system energy management.
- It helps to meet the demand and manages growth better. Precise load forecasting helps the electric utility to make important decisions on purchasing and generating electric power, planning unit commitment to reduce spinning reserve capacity, and scheduling maintenance plans, network planning and infrastructure development.
- Besides playing a key role in reducing the generation cost, it is also essential for the reliability of the power system.
- The power system operator uses the load forecasting results to prepare for load shedding, power purchase, hydro scheduling, hydro-thermal coordination and switching on/off peaking units.
- Load forecasts are important for energy suppliers, independent system operators, financial institutions, and other participants in electric energy generation, transmission, distribution, and energy markets.

4.2 LSTM-based Load Forecasting Framework

- The LSTM model has five crucial elements: input gate (it), forget gate (ft), output gate (Ot), update signal (U) and state output as shown in figure 4.1.
- These gates operate as reading, writing, and erasing for cell memory states (Kong et al., 2018c). The input data of LSTM at time t is xt , the output value is ht, Ct is the memory state, and h(t−1) is the output of the previous cell (Jiao et al., 2018).
- The framework starts from collection of ESM data from cloud (Microsoft Azure) then pre-process the input data such as smoothing and scaling in range 0 to 1 using pre-processing techniques (Singh, 2019), which is then fed into the LSTM block. LSTM memory cell is updated as follows:

$$f_t = \sigma(w_f[h_{(t-1)}, x_t] + b_f)$$

$$i_t = \sigma(w_i[h_{(t-1)}, x_t] + b_i)$$

$$U = tanh(w_c[h_{(t-1)}, x_t] + b_c)$$

$$C_t = f_t * C_{(t-1)} + i_t * U$$

$$O_t = \sigma(w_o[h_{(t-1)}, x_t] + b_o)$$

$$h_t = o_t * tanh(C_t)$$

Where wf ,wi,wc,wo are the weight matrices of forget gate, input gate, candidate state, and output gate respectively. bf , bi, bc, bo are bias vectors. There are two activation functions used in LSTM: (sigmoid) and tanh.

The output of the top LSTM layer is energy consumption forecast of the target time interval.

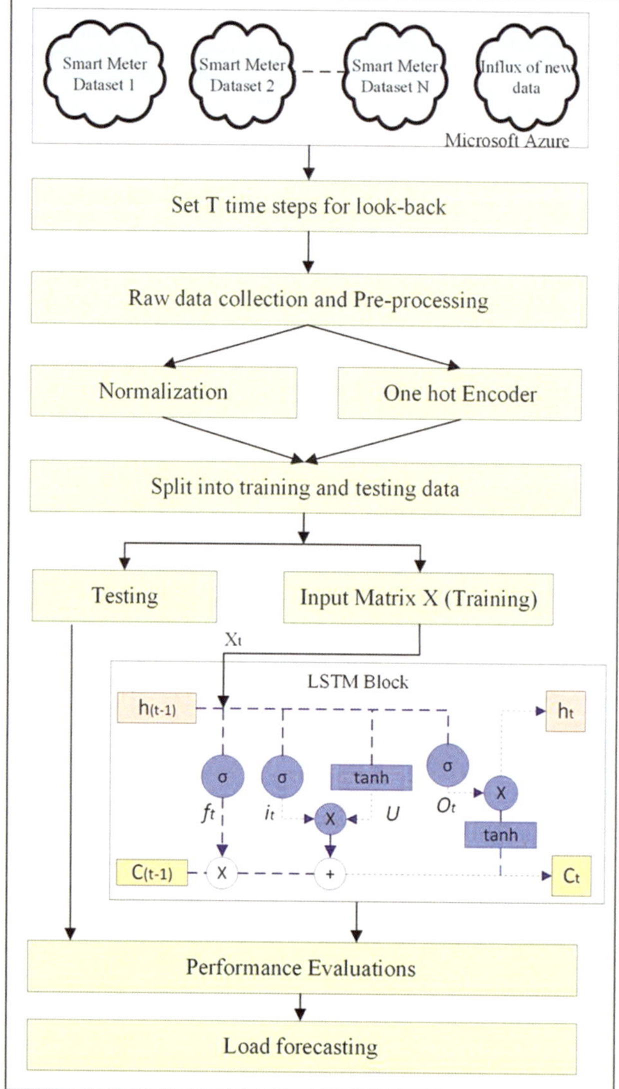

Fig. 4.1. LSTM-based Residential Load Forecasting Framework

(Source: Kong et al. (2018c))

4.3 Experimental Results

The LSTM framework is tested on a real-world smart meter dataset from Toronto, Canada, for forecasting the consumption at a given time interval.

4.3.1 Pre-Processing

- The smart meter data is time-series data and literature survey (Tureczek et al., 2018, 2019) demonstrates the potential existence of autocorrelation in smart meter energy data.
- The dataset used in this research contains a time-dependent component. This component governs information about how previous consumption affects current consumption and exhibits lag in prediction.
- In a paper, Dixit et al., 2015 used Discrete Wavelet Transform (DWT) to get rid of the time lag problem. Hence, we used DWT to scale and translate smart meter data. For the Wavelet analysis, PyWavelet 0.5.2 (Lee et al., 2019) is utilized.
- The wavelet coefficients calculated through the DWT are input to the LSTM model. An important property of the resulting wavelet coefficients is they are uncorrelated (Nason, 2008).

4.3.2 Load Forecasting using LSTM Model: Case Study 1

Datasets Description:

This dataset, provided by the Toronto Hydro-Electric System Ltd, is the electric load data of Toronto (Canada) from 1/1/2016 to 31/7/2016 (Toronto, 1998). The measured electrical data is taken by the hour, so it contains a total of 75,279 samples. Table 4.1 describe the electricity load dataset.

Electricity load in Toronto	
Variables	**Description**
Country	Toronto (Canada)
Supplier	Toronto Hydro Electric System Ltd
No. of Household	1
Monitoring period	1/2016 to 7/2016
Resolution	1 hour
No. of Instances	75,279
Attributes	Date, Time, global active power (kW), global reactive power, voltage

Table 4.1 : Toronto Electricity Load Dataset

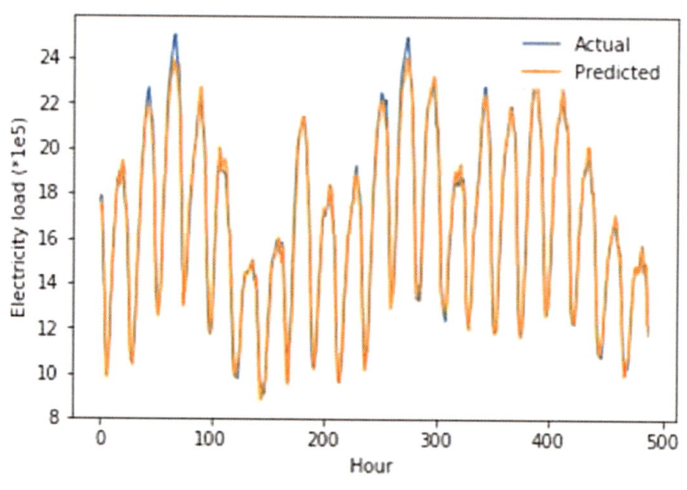

Fig. 4.2. Hour-level Load Forecasting using LSTM Model 

Mean Squared Error	0. 020830

The hour level forecast plots are shown in figure 4.2 and 4.3. The prior knowledge about the forecast can help reshape the load and cut the electricity demand curve, thus allowing a better management and distribution of the electricity in smart grid systems.

MSE is the average of the squared difference between predicted and the actual data points.

Load Forecasting using LSTM Model: Case Study 2

Dataset Description:

The LSTM framework is tested on more real-time European Network of Transmission System Operators for Electricity (ENTSO-E) (Hirth et al., 2018) and National Centers for Environmental Information (NCEI) Integrated Surface Dataset (ISD) (Smith et al., 2011) datasets. This dataset contains 210760 hourly measurements of demand across 9000 customers in kW. Table 4.2 describe the ENTSO-E dataset.

ENTSO-E	
Variables	**Description**
Country	European Countries
Supplier	Transmission System Operators
No. of Customers	9,000
Monitoring period	31/12/2014 to 16/5/2017
Resolution	1 hour
No. of Instances	210,760
Attributes	Electricity consumption, Weather and Calendar data

Table 4.2. ENTSO-E Electricity Load Dataset

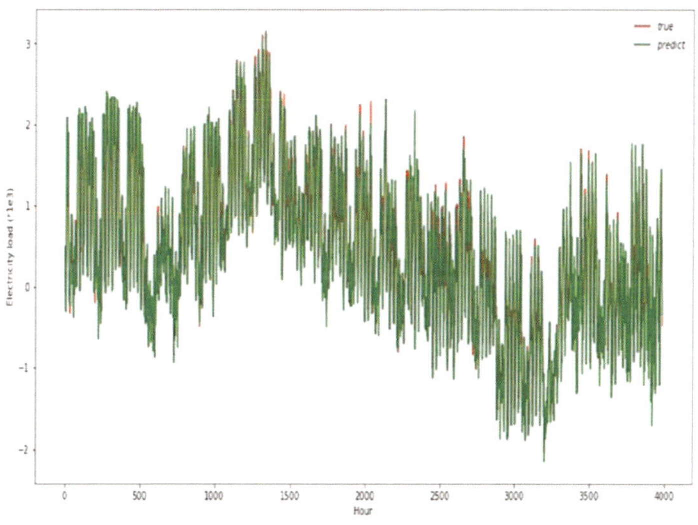

Fig. 4.3. Load Forecasting

MSE is the average of the squared difference between predicted and the actual data points.

The MSE value is found to be minimum which indicates higher estimation accuracy.

Mean Squared Error	0.05490

4.4 Summary of Chapter 4

- Load forecasting for a single customer is hard due to the volatile nature of the individual.
- The LSTM model learns valuable long-distance dependencies in sequence so that it can discover the energy consumption patterns of individual residential consumers.
- From the implementation result, the LSTM module makes accurate load forecasts by exploiting long-term dependencies.
- The performance of the LSTM model was experimentally validated by the real-world data from the Toronto electrical smart meter dataset.
- The MSE value for all two case studies is found to be minimum, which indicates higher estimation accuracy.
- Prior knowledge about the forecast can help reshape the load and cut the electricity demand curve, thus allowing better management and distribution of the electricity in smart grid systems.

Chapter 5

Concluding Remarks and Future Research Directions

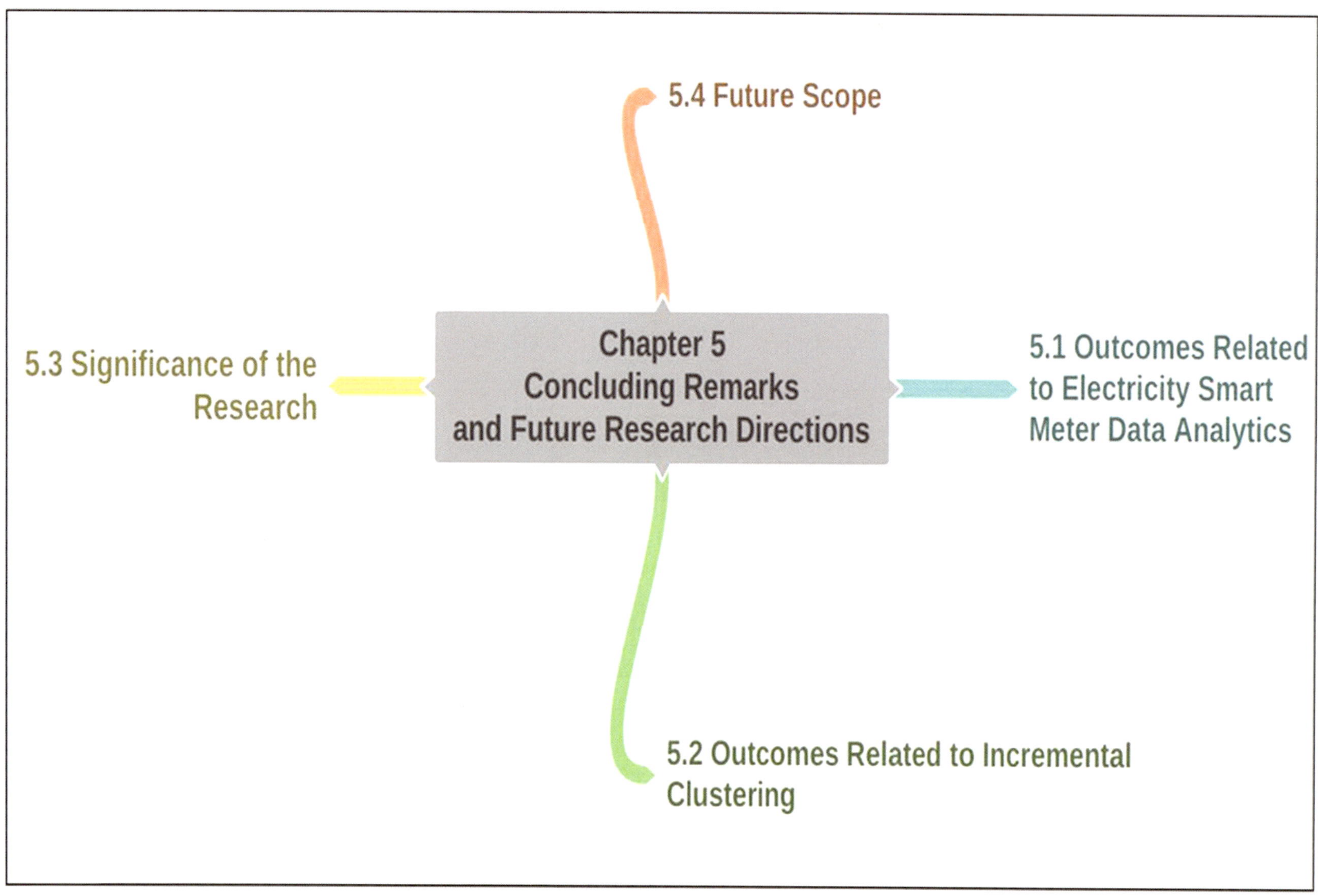

Mind map of Chapter 5

Source: https://coggle.it/

Discussion on the Research Outcomes

- The explosive growth of Electricity Smart Meter (ESM) data and deployments has presented key challenges across the utility industry. ESM effectually communicates with utility companies for monitoring and management of electricity usage as well as with customers for observing electricity consumption.
- ESM generates electricity data dynamically and in real-time mode. Such huge data has embedded patterns and hidden information to extract and learn from. This learning is incremental in nature for all involved entities and users, as the data is growing exponentially in real time.
- To achieve learning from such dynamic data sources, incremental clustering algorithms are mandatory.
- The system presented here, Cloud for Closeness-based Gaussian Mixture Incremental Clustering (Cloud4CGMIC), efficiently handles the incremental ESM data and extracts the required knowledge or hidden insights to improve electricity management for both utilities and customers.
- The system presented here not only assists in lower electricity consumption but also helps reduce carbon emission.
- The reduction of electricity consumption and carbon emission require accurate understanding of hidden patterns of data and related socioeconomic characteristics.

5.1 Outcomes Related to ESM Data Analytics

- As the system presented here is parameter-free, distributed and iterative in nature, it learns automatically from the empirical data about electricity consumption, consumer behaviour, load generation, load profiling, etc.
- Predictions related to electrical energy usage, wastage, outage (both planned and unplanned), etc. are generated for both customers and generation side entities.
- All types of energy sources data can be accommodated in this system, including solar, hydra, conventional and floating solar.
- Farmers in remote areas and villages should get quality and uninterrupted electricity supply to grow quality food and consumables. This system will help understand the electricity requirements of all remote places so that generation units can supply uninterrupted power to offer farmers a good quality of life.
- Finally, the ultimate aim of the varied analysis given by this system includes reduced carbon emissions.
- All these analytical results and predictions will be useful for future researchers not only in the IT domain but also in the energy sector.

5.2 Outcomes Related to Incremental Clustering

- Incremental learning from new labeled or unlabeled instances of data, without discarding previously acquired knowledge
- Formulation of knowledge effectively with any new learning that has occurred
- Evolving with new scenarios as per the problem requirement
- Parameter-free, i.e., free from selecting a number of cluster initialization that may hamper the quality of clusters
- Cluster formation first, unlike other clustering algorithms, which are centroid first
- Cluster ranking during iterations for outlier detection
- Incremental learning via incremental clustering achieved using this system on the arrival of new data in real-time mode, enabling incremental learning about electricity energy generation, utilization, consumption, etc.
- Incremental learning achieved for automatically suggesting groups of clients for specific actions, such as commercial offers for energy reduction
- Design of multifaceted, complete cloud-based distributed incremental clustering; learning and forecasting semi-supervised systems accessible to all stakeholders worldwide

5.3 Social Relevance of Research

The overview of the proposed system is depicted in figure 5.1.

- Household customers will be able to monitor and improvise electricity consumption patterns.
- Utility providers will be able to reduce power outage and avoid capital expenses of building new plants.
- The environment will be protected by reducing pollution via carbon production by power plants (as pollution is hazardous to health).
- The knowledge obtained by the proposed approach can help utilities to find suitable customer groups for specific energy efficiency programs, offer directed electricity savings advice, improve the forecasting accuracy and estimate the electricity consumption pattern of new customers.

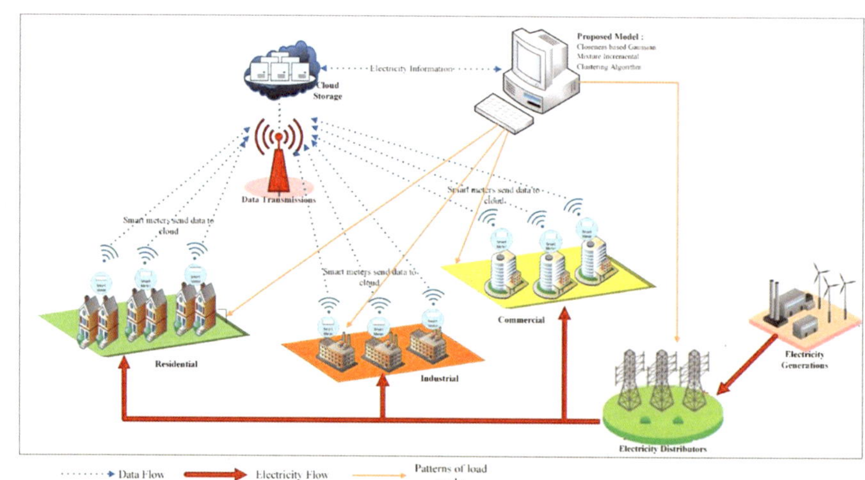

Fig. 5.1. Electricity Smart Meter Data Analysis using Incremental Clustering Algorithm

5.4 Future Scope

- The Distributed Incremental Clustering system presented in this "Quick Reference Book" is the generalized solution applicable to any smart meter data analysis including gas, water, solar, floating solar, etc.
- Energy captured by floating solar plants and data analysis, pattern recognition, and forecasting can be analysed using the system presented in this book in upcoming phases of research.
- The system presented here can be further combined with other machine learning techniques (e.g., K-Nearest Neighbour, NB etc.)
- The generation-side energy requirements, predictions / forecasting, hidden pattern analysis, etc. also can be analysed using this system.
- Data analysis of alternate electricity sources, e.g., generators and inverters, can also be conducted using this system.

Appendix A

patent & paper Publications

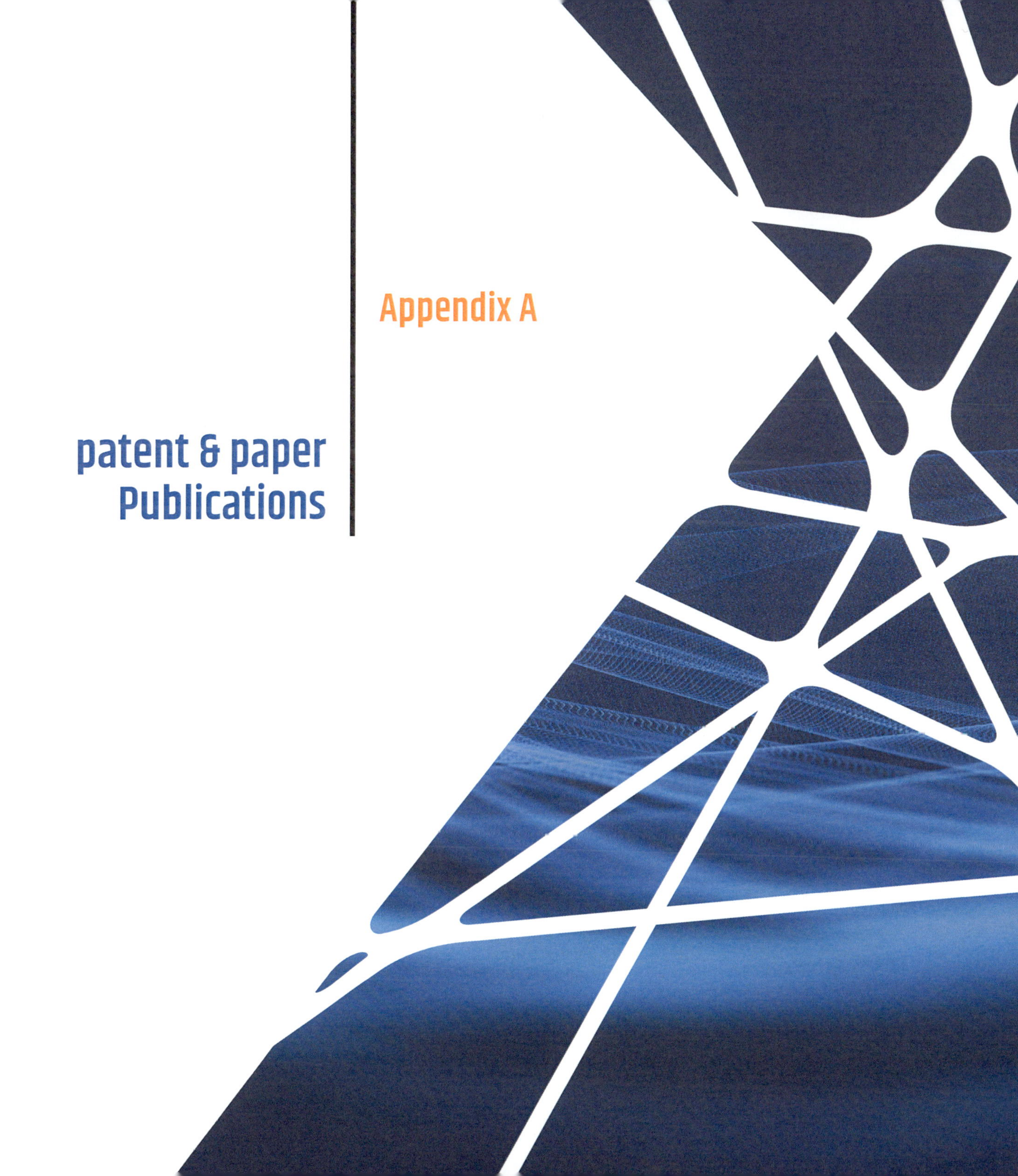

International Journal of Information Retrieval Research
Volume 10 • Issue 2 • April-June 2020

Cloud4NFICA-Nearness Factor-Based Incremental Clustering Algorithm Using Microsoft Azure for the Analysis of Intelligent Meter Data

Archana Yashodip Chaudhari, Symbiosis Institute of Technology, Symbiosis International (Deemed University), Pune, India
Preeti Mulay, Symbiosis Institute of Technology, Symbiosis International (Deemed University), Pune, India
https://orcid.org/0000-0002-4779-6726

Science & Technology Libraries

Routledge

Distributed Incremental Clustering Algorithms: A Bibliometric and Word-Cloud Review Analysis

Preeti Mulay, Rahul Joshi & Archana Chaudhari

To cite this article: Preeti Mulay, Rahul Joshi & Archana Chaudhari (2020) Distributed Incremental Clustering Algorithms: A Bibliometric and Word-Cloud Review Analysis, Science & Technology Libraries, DOI: 10.1080/0194262X.2020.1775163

To link to this article: https://doi.org/10.1080/0194262X.2020.1775163

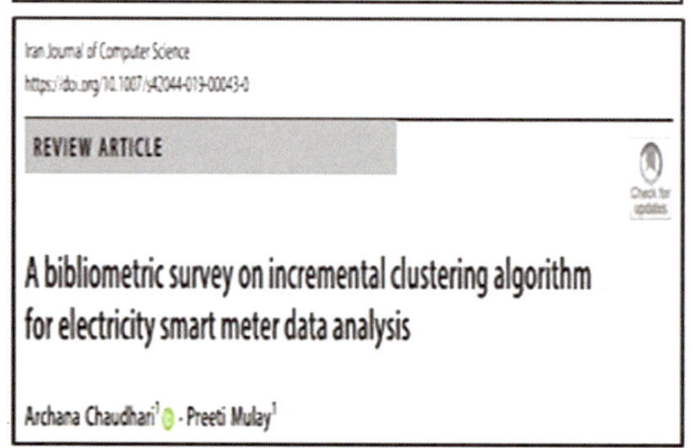

Iran Journal of Computer Science
https://doi.org/10.1007/s42044-019-00043-0

REVIEW ARTICLE

A bibliometric survey on incremental clustering algorithm for electricity smart meter data analysis

Archana Chaudhari[1] · Preeti Mulay[1]

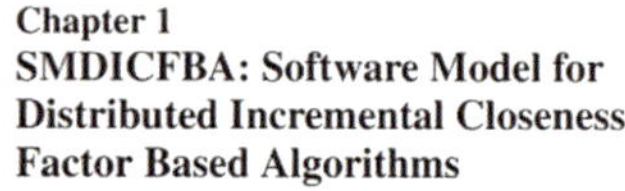

Chapter 1
SMDICFBA: Software Model for Distributed Incremental Closeness Factor Based Algorithms

Rahul Raghvendra Joshi, Preeti Mulay and Archana Chaudhari

International Journal of Modern Agriculture, Volume 9, No.3, 2020
ISSN: 2305-7246

Mobile Phone Charging: Power Statistics & Energy Consumption Pattern Analysis Using Developed "Powerstats" Android Application

Saloni Kuralkar[1], Preeti Mulay[2], Archana Chaudhari[3]
[1,2,3]Symbiosis Institute of Technology, Symbiosis International (Deemed University), Pune, India
Email: [2]preeti.mulay@sitpune.edu.in

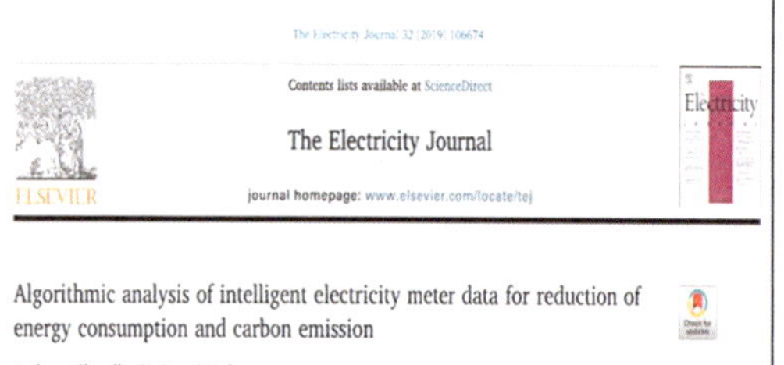

Contents lists available at ScienceDirect

The Electricity Journal

journal homepage: www.elsevier.com/locate/tej

Algorithmic analysis of intelligent electricity meter data for reduction of energy consumption and carbon emission

Archana Chaudhari*, Preeti Mulay

International Journal of Modern Agriculture, Volume 9, No.3, 2020
ISSN: 2305-7246

Mobile Phone Charging: Power Statistics & Energy Consumption Pattern Analysis Using Developed "Powerstats" Android Application

Saloni Kuralkar[1], Preeti Mulay[2], Archana Chaudhari[3]
[1,2,3]Symbiosis Institute of Technology, Symbiosis International (Deemed University), Pune, India
Email: [2]preeti.mulay@sitpune.edu.in

The current issue and full text archive of this journal is available on Emerald Insight at: https://www.emerald.com/insight/1750-6220.htm

Unleashing analytics to reduce electricity consumption using incremental clustering algorithm

Analytics to reduce electricity consumption

Archana Yashodip Chaudhari
Department of Computer Engineering, Dr. D. Y. Patil Institute of Technology, Pimpri, Pune and Savitribai Phule Pune University, Pune, India, and
Preeti Mulay
Symbiosis Institute of Technology, Symbiosis International (Deemed University), Pune, India

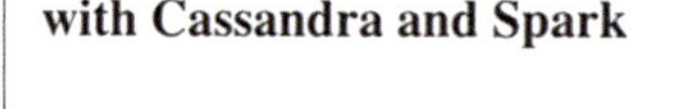

SCSI: Real-Time Data Analysis with Cassandra and Spark

Archana A. Chaudhari and Preeti Mulay

University of Nebraska - Lincoln
DigitalCommons@University of Nebraska - Lincoln

Library Philosophy and Practice (e-journal) — Libraries at University of Nebraska-Lincoln

Fall 9-16-2019

Bibliometric Survey on Incremental Clustering Algorithms

Archana Chaudhari
Research Scholar, Symbiosis Institute of Technology (SIT) affiliated to Symbiosis International (Deemed University), Pune, India, archana.chaudhari@sitpune.edu.in

Rahul Raghvendra Joshi
Research Scholar and Assistant Professor, Symbiosis Institute of Technology (SIT) affiliated to Symbiosis International (Deemed University), Pune, India, rahulj@sitpune.edu.in

Preeti Mulay
Ph.D. Guide and Associate Professor, Symbiosis Institute of Technology (SIT) affiliated to Symbiosis International (Deemed University), Pune, India, preeti.mulay@sitpune.edu.in

Ketan Kotecha
Director, Symbiosis Institute of Technology (SIT) affiliated to Symbiosis International (Deemed University), Pune, India, director@sitpune.edu.in

Parag Kulkarni
Innovation Strategist, CEO, and Chief Scientist, iknowlation Research Labs Pvt Ltd Pune, India, paragindia@gmail.com

Follow this and additional works at: https://digitalcommons.unl.edu/libphilprac
Part of the Computer Engineering Commons, and the Library and Information Science Commons

Science & Technology Libraries

Routledge

Smart Energy Meter: Applications, Bibliometric Reviews and Future Research Directions

Saloni Kuralkar, Preeti Mulay & Archana Chaudhari

University of Nebraska - Lincoln
DigitalCommons@University of Nebraska - Lincoln

Library Philosophy and Practice (e-journal) — Libraries at University of Nebraska-Lincoln

Summer 5-7-2020

Bibliometric Study of Bibliometric Papers about Clustering

Preeti Mulay
Symbiosis Institute of Technology (SIT), Symbiosis International (Deemed University) (SIU), Pune, India, preeti.mulay@sitpune.edu.in

Rahul Raghvendra Joshi
Symbiosis Institute of Technology (SIT), Symbiosis International (Deemed University) (SIU), Pune, India, rahulj@sitpune.edu.in

Archana Chaudhari
Symbiosis Institute of Technology (SIT), Symbiosis International (Deemed University) (SIU), Pune, India, archana.chaudhari@sitpune.edu.in

Research Papers Published till Date

1	"Cloud4NFICA- Nearness Factor based Incremental Clustering Algorithm using Microsoft Azure for Analysis of Intelligent Meter Data", International Journal of Information Retrieval Research, vol. 10, no. 2, pp.21-39, 2020. [WoS] http://doi.org/10.4018/IJIRR.2020040102
2	"Algorithmic Analysis of Intelligent Electricity Meter Data for Reduction of Energy Consumption and Carbon Emission ", The Electricity Journal, vol. 32, no. 10, pp.1-9, 2019. [Elsevier, Scopus, Q1] https://doi.org/10.1016/j.tej.2019.106674
3	"Bibliometric Survey on Incremental Clustering Algorithms", Library Philosophy and Practice (e-journal), 2762, 2019. [Scopus, Q2] https://digitalcommons.unl.edu/libphilprac/2762/
4	"A bibliometric survey on incremental clustering algorithm for electricity smart meter data analysis", Iran Journal of Comput Science, 2019. [Springer, EBSCO Discovery Service] https://doi.org/10.1007/s42044-019-00043-0
5	"Smart Energy Meter: Applications, Bibliometric Reviews and Future Research Directions", Science & Technology Libraries, pp 1-24, 2020. [Scopus, Q2] https://doi.org/10.1080/0194262X.2020.1750081
6	"Bibliometric study of Bibliometric Papers about Clustering", Library Philosophy and Practice [Scopus, Q2] https://digitalcommons.unl.edu/libphilprac/4211/

7	“Distributed Incremental Clustering Algorithms: A Bibliometric &Word-Cloud Review Analysis”, Science & Technology Libraries Journal [Scopus, Q2] https://doi.org/10.1080/0194262X.2020.1775163
8	“Unleashing Analytics to Reduce Electricity Consumption using Incremental Clustering Algorithm ”, International Journal of Energy Sector Management.[Scopus, ESCI, ABDC] https://doi.org/10.1108/IJESM-11-2019-0016
9	“Mobile Phone Charging: Power Statistics & Energy Consumption Pattern Analysis using Developed Powerstat android Application”, International Journal of Modern Agriculture, vol 9, no. 3, pp. 1682-1710, 2020 [ESCI] http://modern-journals.com/index.php/ijma/article/view/353
10	“SCSI: Real-Time Data Analysis with Cassandra and Spark”, In: Mittal M., Balas V., Goyal L., Kumar R. (eds) Big Data Processing Using Spark in Cloud. Studies in Big Data, vol 43. Springer, Singapore, 2019, pp 237-264 [ISI Web of Science] https://doi.org/10.1007/978-981-13-0550-4_11
11	“SMDICFBA: Software Model for Distributed Incremental Closeness Factor Based Algorithms”, In: Singh, J., Bilgaiyan, S., Mishra, B.S.P., Dehuri, S. (Eds.) A Journey Towards Bio-inspired Techniques in Software Engineering. Intelligent Systems Reference Library, Springer Nature Switzerland AG, 2020. [Scopus, ISI Web of Science] DOI:10.1007/978-3-030-40928-9_1
12	"Mapping of Six Sigma to Threshold Based Incremental Clustering Algorithm," 2018 IEEE Punecon, Pune, India, 2018, pp. 1-8. [Scopus] https://ieeexplore.ieee.org/document/8745432
13	“Analysis of Chinese Patents associated with Incremental Clustering Algorithms”, Journal of Computing Research and Innovation, vol 7, no. 1, 2022. [UGC Journal] https://doi.org/10.24191/jcrinn.v7i1.266
14	“Text Document Learning using Distributed Incremental Clustering Algorithm: Educational Certificates”, Int. J. of Business Intelligence and Data Mining. [Accepted, Scopus]

Patent Publications

(12) INTERNATIONAL APPLICATION PUBLISHED UNDER THE PATENT COOPERATION TREATY (PCT)

(19) World Intellectual Property Organization
International Bureau

(43) International Publication Date
19 March 2020 (19.03.2020)

WIPO | PCT

(10) International Publication Number
WO 2020/053846 A2

(51) **International Patent Classification:**
G01R 21/133 (2006.01) *G05B 13/04* (2006.01)
G05B 19/048 (2006.01)

(21) **International Application Number:** PCT/IB2020/050108

(22) **International Filing Date:** 08 January 2020 (08.01.2020)

(25) **Filing Language:** English

(26) **Publication Language:** English

(30) **Priority Data:**
201921047841 22 November 2019 (22.11.2019) IN

(71) **Applicant: SYMBIOSIS INTERNATIONAL (DEEMED UNIVERSITY)** [IN/IN]; Gram: Lavale, Taluka Mulshi, Pune, Maharashtra 412115 (IN).

(72) **Inventors: CHAUDHARI, Archana Atmaram**, Symbiosis International (Deemed University) Gram: Lavale, Taluka Mulshi, Pune, Maharashtra 412115 (IN). **MULAY, Preeti Milind**, Symbiosis International (Deemed University) Gram: Lavale, Taluka Mulshi, Pune, Maharashtra 412115 (IN).

(74) **Agent: GUPTA, Priyank**, Stratjuris Law Partners, Office 203 & 204, Supreme Headquarters, Above Tata Motors Showroom, Mumbai - Pune Highway, Mohan Nagar, Baner, Pune, Maharashtra 411045 (IN).

(81) **Designated States** *(unless otherwise indicated, for every kind of national protection available)*: AE, AG, AL, AM, AO, AT, AU, AZ, BA, BB, BG, BH, BN, BR, BW, BY, BZ, CA, CH, CL, CN, CO, CR, CU, CZ, DE, DJ, DK, DM, DO, DZ, EC, EE, EG, ES, FI, GB, GD, GE, GH, GM, GT, HN, HR, HU, ID, IL, IN, IR, IS, JO, JP, KE, KG, KH, KN, KP, KR, KW, KZ, LA, LC, LK, LR, LS, LU, LY, MA, MD, ME, MG, MK, MN, MW, MX, MY, MZ, NA, NG, NI, NO, NZ, OM, PA, PE, PG, PH, PL, PT, QA, RO, RS, RU, RW, SA, SC, SD, SE, SG, SK, SL, ST, SV, SY, TH, TJ, TM, TN, TR, TT, TZ, UA, UG, US, UZ, VC, VN, WS, ZA, ZM, ZW.

1. PCT Publication WO 2020/053846 A2: "A system and method for analysis of smart meter data", filed on January 8, 2020. [Patent Published on 19th March 2020]
2. Indian Patent 201921047841: "A system and method for analysis of smart meter data" , Provisional filed November 22, 2019. [Patent Published on 20th Dec 2019]

References

- Ang, H. H., Gopalkrishnan, V., Hoi, S. C., and Ng, W. K. (2013). Classification in p2p networks with cascade support vector machines. ACM Transactions on Knowledge Discovery from Data (TKDD), 7(4):1–29.
- Arbelaitz, O., Gurrutxaga, I., Muguerza, J., P´erez, J. M., and Perona, I. (2013). An extensive comparative study of cluster validity indices. Pattern Recognition, 46(1):243–256.
- Bao, J., Wang, W., Yang, T., and Wu, G. (2018). An incremental clustering method based on the boundary profile. PLoS ONE, 13(4):1–10.
- Blaabjerg, F., Teodorescu, R., Liserre, M., and Timbus, A. V. (2006). Overview of control and grid synchronization for distributed power generation systems. IEEE Transactions on Industrial Electronics, 53(5):1398–1409.
- Bendechache, M. and Kechadi, M. (2015). Distributed clustering algorithm for spatial data mining. In 2015 2nd IEEE International Conference on Spatial Data Mining and Geographical Knowledge Services (ICSDM), pages 60–65, Fuzhou
- Bhimani, J., Leeser, M., and Mi, N. (2015). Accelerating K-Means clustering with parallel implementations and GPU computing. In 2015 IEEE High Performance Extreme Computing Conference (HPEC), pages 1–6, Waltham, MA, USA. IEEE.
- Barker, S., Mishra, A., Irwin, D., Cecchet, E., Shenoy, P., and Albrecht, J. (2012). Smart*: An open data set and tools for enabling research in sustainable homes. In Workshop on Data Mining Applications in Sustainability (SustKDD 2012). Retrieved from http://traces.cs.umass.edu/index.php/Smart/Smart.
- Chicco, G. (2012). Overview and performance assessment of the clustering methods for electrical load pattern grouping. Energy, 42(1):68–80.
- Chatterjee, N. (1977). FERC:Federal Energy Regulatory Commission. Web page. Retrieved from https://www.ferc.gov/.

- Chaudhari, A., Joshi, R. R., Mulay, P., Kotecha, K., and Kulkarni, P. (2019). Bibliometric survey on incremental clustering algorithms. Library Philosophy and Practice (e-journal), 2762:1–25.
- Chaudhari, A. and Mulay, P. (2019a). A System and Method for Analysis of Smart Meter Data. Indian Patent, (201921047841). Retrieved from http://ipindiaservices.gov.in/PublicSearch/PublicationSearch/PatentDetails.
- Chaudhari, A. and Mulay, P. (2019b). Algorithmic analysis of intelligent electricity meter data for reduction of energy consumption and carbon emission. The Electricity Journal, 32(10):106674.
- Chaudhari, A. and Mulay, P. (2019c). A bibliometric survey on incremental clustering algorithm for electricity smart meter data analysis. Iran Journal of Computer Science, 2(4):197–206.
- Chaudhari, A. and Mulay, P. (2020a). A System and Method for Analysis of Smart Meter Data. Patent Cooperation Treaty (PCT), (WO2020/053846 A2). Retrieved from https : //patentscope.wipo.int/search/en/detail.jsf?docId =WO2020053846 &c id = P12 – K7Y UIR – 21923 – 1
- Chaudhari, A. Y. and Mulay, P. (2020b). Cloud4NFICA-Nearness Factor-Based Incremental Clustering Algorithm Using Microsoft Azure for the Analysis of Intelligent Meter Data. International Journal of Information Retrieval Research (IJIRR), 10(2):21–39.
- Chen, Y.-T. (2017). The factors affecting electricity consumption and the consumption characteristics in the residential sector - a case example of taiwan. Sustainability, 9(8):1484.
- Daftari, S. (2018). Analysis of Smart meter London. GitHub. Retrieved from https : //github.com/Sunanda1/Analysis – of – Smart – meter – London.git.
- Davies, D. L. and Bouldin, D. W. (1979). A Cluster Separation Measure. IEEE Transactions on Pattern Analysis and Machine Intelligence, PAMI-1(2):224 – 227.
- Dixit, P., Londhe, S., and Dandawate, Y. (2015). Removing prediction lag in wave height forecasting using neuro-wavelet modeling technique. Ocean Engineering, 93:74–83.
- Ester, M., Kriegel, H.-P., Sander, J., and Xu, X. (1996). A density-based algorithm for discovering clusters a density-based algorithm for discovering clusters in large spatial databases with noise. In Proceedings of the Second International Conference on Knowledge Discovery and Data Mining, KDD'96, pages 226 – 231. AAAI Press.
- EPRI (2020). Electric Power Research Institutes IntelliGrid SM Initiative. Web page. Retrieved from http://intelligrid.epri.com/.
- Flath, C., Nicolay, D., Conte, T., Dinther, C., and Filipova-Neumann, L. (2012). Cluster Analysis of Smart Metering Data:An Implementation in Practice. Business & Information Systems Engineering, 4:31–39.
- Fisher, D. H. (1987). Knowledge acquisition via incremental conceptual clustering.Machine learning, 2(2):139–172.
- Fallah, S., Deo, R., Shojafar, M., Conti, M., and Shamshirband, S. (2018). Computational intelligence approaches for energy load forecasting in smart energy management grids: State of the art, future challenges, and research directions. Energies, 11(596):1–31.
- Granell, R., Axon, C. J., and Wallom, D. C. (2015). Impacts of raw data temporal resolution using selected clustering methods on residential electricity load profiles. IEEE Transactions on Power Systems, 30(6):3217–3224.

- Huang, F., Zhu, Q., Zhou, J., Tao, J., Zhou, X., Jin, D., Tan, X., and Wang, L. (2017). Research on the parallelization of the DBSCAN clustering algorithm for spatial data mining based on the spark platform. Remote Sensing, 9(12):1301.
- Hirth, L., M¨uhlenpfordt, J., and Bulkeley, M. (2018). The entso-e transparency platform – a review of Europe's most ambitious electricity data platform. Applied Energy, 225:1054–1067.
- ISSDA (14 April 2018). Irish Social Science Data Archive: Commission for Energy Regulation (CER) smart metering project. Retrieved from http://www.ucd.ie/issda/data/commissionforenergyregulationcer/.
- Joshi, P. and Kulkarni, P. (2012). Incremental learning: Areas and methods-a survey. International Journal of Data Mining & Knowledge Management Process, 2(5):43–51.
- Jiang, Y., Liu, C.-C., Diedesch, M., Lee, E., and Srivastava, A. K. (2015). Outage management of distribution systems incorporating information from smart meters. IEEE Transactions on Power Systems, 31(5):4144–4154.
- Jiao, R., Zhang, T., Jiang, Y., and He, H. (2018). Short-term non-residential load forecasting based on multiple sequences lstm recurrent neural network. IEEE Access, 6:59438–59448.
- Kotary, D. K. and Nanda, S. J. (2020). Distributed robust data clustering in wireless sensor networks using diffusion moth flame optimization. Engineering Applications of Artificial Intelligence, 87:103342.
- Kulkarni, P. and Joshi, P. (2015). Artificial intelligence: building intelligent systems. Prentice Hall India Learning Private Limited, Delhi.
- Kulkarni, P. A. and Mulay, P. (2013). Evolve systems using incremental clustering approach. Evolving Systems, 4(2):71–85.
- Kong, W., Dong, Z. Y., J.Hill, D., Luo, F., and Xu, Y. (2018b). Short-term residential load forecasting based on resident behaviour learning. IEEE Transactions on Power Systems, 33(1):1087–1088.
- Kong, W., Dong, Z. Y., Jia, Y., Hill, D. J., Xu, Y., and Zhang, Y. (2018c). Short-term residential load forecasting based on lstm recurrent neural network. IEEE Transactions on Smart Grid, 8(1):1–11.
- Lee, G., Gommers, R., Wasclcwski, F., Wohlfahrt, K., and O'Leary, A. (2019). PyWavelets : A Python package for wavelet analysis. Journal of Open Source Software, 4(36):1237.
- Li, Z. and Yao, T. (2010). Renewable Energy Basing on Smart Grid. In 2010 6th International Conference on Wireless Communications Networking and Mobile Computing (WiCOM), pages 1–4, Chengdu. IEEE.
- Li, K., Ma, Z., Robinson, D., and Ma, J. (2018). Identification of typical building daily electricity usage profiles using gaussian mixture model-based clustering and hierarchical clustering. Applied energy, 231:331–342.
- MSEB (22 May 2018). SCADA Data: Maharashtra Generation, Exchange and Demand overview and Load shedding data. Retrieved from http://beta.mahadiscom.in/scheduled-outage-informationnov-2016/.
- Ma, Z., Xie, J., Li, H., Sun, Q., Si, Z., Zhang, J., and Guo, J. (2017). The role of data analysis in the development of intelligent energy networks. IEEE Network, 31(5):88–95.
- Mammen, P. M., Kumar, H., Ramamritham, K., and Rashid, H. (2018). Want to Reduce Energy Consumption, Whom Should We Call? In Proceedings

of the Ninth International Conference on Future Energy Systems, e-Energy '18, pages 12–20, New York, NY, USA. Association for Computing Machinery.

- Mrozek, D., Gosk, P., and Ma lysiak-Mrozek, B. (2015). Scaling ab initio predictions of 3d protein structures in microsoft azure cloud. Journal of Grid Computing, 13(4):561–585.
- Nason, G. (2008). Wavelet methods in statistics with R. Springer Science & Business Media.
- OPSD: Open Power System Data (2020). A Free and Open Data Platform for Power System Modelling. Webpage. Retrieved from http://open-power-system-data.org.
- Prayas Energy Group (16 November 2018). Smart meter consumer data (Pune). Retrived from http://www.prayaspune.org/peg/research-areas/electricitygeneration- supply.html.
- Pawar, S. and Momin, B. (2017). Smart electricity meter data analytics: A brief review. In 2017 IEEE Region 10 Symposium (TENSYMP), pages 1–5, Cochin. IEEE.
- Pabon, M., Eveleigh, T., and Tanju, B. (2017). Smart meter data analytics for optimal customer selection in demand response programs. 107:49–59.
- Predd, J. B., Kulkarni, S. B., and Poor, H. V. (2006). Distributed learning in wireless sensor networks. IEEE Signal Processing Magazine, 23(4):56–69.
- Pfander, D., Daiß, G., and Pfl¨uger, D. (2019). Heterogeneous Distributed Big Data Clustering on Sparse Grids. Algorithms, 12(3):1–21.
- Qin, J., Fu, W., Gao, H., and Zheng, W. X. (2017). Distributed k-means algorithm and fuzzy c-means algorithm for sensor networks based on multiagent consensus theory. IEEE transactions on cybernetics, 47(3):772–783.
- Rashid, H., Singh, P., and Singh, A. (2019). I-blend, a campus-scale commercial and residential buildings electrical energy dataset. Scientific Data, 6(190015):1–12.
- S. Chakraborty and N. K. Nagwani (2011) "Analysis and Study of Incremental K-Means Clustering Algorithm," Berlin, Heidelberg, vol 10, no.2 pp. 338-341.
- Schofield, J. R., Carmichael, R., Tindemans, S. H., Bilton, M., Woolf, M., and Strbac, G. (2013). Low carbon london project; data from the dynamic time-of-use electricity pricing trial. Retrived from https://beta.ukdataservice.ac.uk/datacatalogue/doi/?id=7857-1.
- Siano, P. (2014). Demand response and smart grids - a survey. Renewable and sustainable energy reviews, 30:461–478.
- Singh, P. (2019). Mllib: Machine learning library. In Learn PySpark, pages 85–115. Springer.
- Shin, C., Lee, E., Han, J., Yim, J., Rhee, W., and Lee, H. (2019). The enertalk dataset, 15 hz electricity consumption data from 22 houses in korea. Scientific Data, 6(193):1–13.
- Silva, J. A., Faria, E. R., Barros, R. C., Hruschka, E. R., Carvalho, A. C. d., and Gama, J. (2013). Data stream clustering: A survey. ACM Computing Surveys (CSUR), 46(1):1–31.

- Smith, A., Lott, N., and Vose, R. (2011). The integrated surface database: Recent developments and partnerships. Bulletin of the American Meteorological Society, 92(6):704–708.
- Toronto, M. G. (1998). Toronto Hydro Corporation. Web Page. Retrieved from https://www.torontohydro.com/SITES/ELECTRICSYSTEM/BUSINESS/.
- Tureczek, A., Nielsen, P. S., and Madsen, H. (2018). Electricity consumption clustering using smart meter data. Energies, 11(4):859 – 877.
- Tureczek, A. M., Nielsen, P. S., Madsen, H., and Brun, A. (2019). Clustering district heat exchange stations using smart meter consumption data. Energy and Buildings, 182:144–158.
- Visalakshi, N. K. and Thangavel, K. (2009). Distributed data clustering: a comparative analysis. In Abraham, A., Hassanien, A.-E., de Leon F. de Carvalho, A. P., and Sn´aˇsel, V., editors, Foundations of Computational, Intelligence, volume 6, pages 371–397. Springer Berlin Heidelberg, Berlin, Heidelberg.
- Wang, Y., Chen, Q., Hong, T., and Kang, C. (2018). Review of smart meter data analytics: Applications, methodologies, and challenges. IEEE Transactions on Smart Grid, 10(3):3125–3148.
- Walker, J. (2020). The data set: Open power systems data. Tutorial: Time Series Analysis with Pandas. Retrieved from https://www.dataquest.io/blog/tutorialtime-series-analysis-with-pandas/.
- Wiese, F., Schlecht, I., Bunke, W.-D., Gerbaulet, C., Hirth, L., Jahn, M., Kunz, F., Lorenz, C., M¨uhlenpfordt, J., Reimann, J., et al. (2019). Open power system data– frictionless data for electricity system modelling. Applied energy, 236:401–409.
- Yildiz, B., Bilbao, J., Dore, J., and Sproul, A. (2017). Recent advances in the analysis of residential electricity consumption and applications of smart meter data. Applied Energy, 208:402 – 427.
- Zhou, J., Chen, L., Chen, C. P., Wang, Y., and Li, H.-X. (2017). Uncertain data clustering in distributed peer-to-peer networks. IEEE transactions on neural networks and learning systems, 29(6):2392–2406.

About the Authors

Dr. Archana Y. Chaudhari

- Dr. Archana Chaudhari earned PhD in Computer Engineering from Symbiosis International (Deemed University), Pune.
- Her areas of research focuses on Energy, Smart Electricity Meter data analysis, outage management, machine learning and incremental learning.
- She is the recipient of grant from "Sakal India Foundation", "AI for Earth" from Microsoft Azure for her energy and carbon emission related research and Best Poster Award at KPIT-IISER, Pune Energy & Mobility PhD Conference in 2022.
- She is reviewer of International Journals and Conferences. She Published research papers in Elsevier, Springer, InderScience, Emerald and IEEE Xplore.
- She is currently working in Computer Engineering Department of Dr. D. Y. Patil Institute of Technology, Pimpri, Pune.

Dr. Preeti Mulay

- Dr. Preeti Mulay is working as Professor in the department of Computer Science with Symbiosis Institute of Technology, Pune, India.
- She has authored or coauthored many technical papers in various journals and conferences.
- Her areas of interest include machine learning, data mining, software engineering and knowledge augmentation.
- She received Microsoft Azure's "Data Science" research grant for the integrated research related to Machine Learning and Diabetes study. She is also the recipient of Microsoft Azure's "AI for Earth" research grants for the research related to applications of Incremental Learning for Electricity Smart-meter data analysis.
- She is recognized as IEEE reviewer and certified Springer Reviewer.

www.ingramcontent.com/pod-product-compliance
Ingram Content Group UK Ltd.
Pitfield, Milton Keynes, MK11 3LW, UK
UKRC032309290726
14090UKWH00004B/410

* 9 7 8 9 3 9 5 1 3 9 5 2 6 *